Performance-Enhancing Medications and Drugs of Abuse

Mark S. Gold, MD

Editor

Performance-Enhancing Medications and Drugs of Abuse has been co-published simultaneously as *Journal of Addictive Diseases*, Volume 25, Supplement Number 1 2007.

52 Vanderbilt Avenue
New York, NY 10017

Telephone House
69-77 Paul Street
London EC2A 4LQ, UK

www.informahealthcare.com

Informa Healthcare USA, Inc.
52 Vanderbilt Avenue
New York, NY 10017

International Standard Book Number-10: 0-7890-3665-7 (Softcover)
International Standard Book Number-13: 978-0-7890-3665-0 (Softcover)

Library of Congress Cataloging-in-Publication Data

Performance-enhancing medications and drugs of abuse / Mark S. Gold, editor.
    p. ; cm. "Co-published simultaneously as Journal of addictive diseases, volume 25, supplement number 1 2007."
  Includes bibliographical references and index.
  ISBN-13: 978-0-7890-3665-0 (softcover : alk. paper)
  ISBN-10: 0-7890-3665-7 (Softcover : alk. paper)
  1. Doping in sports. I. Gold, Mark S.
  [DNLM: 1. Substance-Related Disorders. 2. Doping in Sports. 3. Pharmaceutical Preparations. WM 270 P438 2007]

RC1230.P477 2007
362.29-dc22   2007001985

Visit the Informa Web site at
www.informa.com

and the Informa Healthcare Web site at
www.informahealthcare.com

Performance-Enhancing Medications and Drugs of Abuse

CONTENTS

Introduction   1
Mark S. Gold
Lisa J. Merlo

A Review of Performance-Enhancing Drugs in Professional Sports and Their Spread
to Amateur Athletics, Adolescents, and Other At-Risk Populations   5
David M. Martin
David A. Baron
Mark S. Gold

Nicotine: Risks and Performance-Enhancing Effects   17
C. Patrick Lane
Noni A. Graham
Ellen A. Ovson

Anabolic Steroid Abuse: Neurobiological Substrates and Psychiatric Comorbidity   33
Daria Rylkova
Adrie W. Bruijnzeel
Mark S. Gold

Performance Enhancement and Adverse Consequences of MDMA   47
Firas H. Kobeissy
Meghan B. O'Donoghue
Erin C. Golden
Stephen F. Larner
Zhiqun Zhang
Mark S. Gold

Performance Enhancing, Non-Prescription Use of Erectile Dysfunction Medications   61
Noni A. Graham
Alexandria (Alexis) Polles
Mark S. Gold

Index   69

ABOUT THE EDITOR

Mark S. Gold, MD, is a Distinguished Professor of Psychiatry, Neuroscience, and Community Health & Family Medicine at the University of Florida College of Medicine. He is also a member of the McKnight Brain Institute and is Chief of the Addiction Medicine Division in the Department of Psychiatry. He has written over 900 medical articles, chapters, and abstracts in health professional journals on wide variety of psychiatric research subjects, and authored 12 professional books, including practice guidelines, ASAM core competencies, and medical textbooks for primary care professional. He is also author of 15 general audience books. As a researcher and inventor, he worked for nearly 30 years to develop models for understanding the effects of tobacco and other drugs on the brain and behavior.

Dr. Gold is the mentor of many of the nation's current leaders in eating disorders and addiction education and research. He is active in numerous national organizations, including the National Scientific Advisory Board for Media Partnership for a Drug Free America, DARE, the American Council for Drug Education, and PRIDE.

Introduction

Historically, psychiatrists faced with a new patient have completed a physical and neurological exam, administered a clinical interview, obtained collateral information, and then assigned a diagnosis according to the diagnostic nosology of the day. Today however, many practitioners evaluate their patients based on what the patient reports wanting from the visit: what is the problem or symptom that needs to be solved or reversed? This tendency to treat the symptom, rather than the patient or the disease, has resulted in the devolution of psychiatric care into 10-minute medication-check office visits, checklist diagnoses, and treatment-on-demand with the pharmaceutical of the patients' choice. Unfortunately, it is no great hurdle for patients to identify aspects of their thoughts or behavior that might respond to a medication or drug of abuse and then determine which drugs to seek. Indeed, if one is tired, or excessively shy at parties, or wants to lose weight, or lacks concentration, or wants to putt better on the 18th green, modern cosmetic psychiatry has a solution. Yet, despite the recent proliferation of performance-enhancing drug-seeking, Addiction Medicine and Addiction Psychiatry specialists generally have not received referrals related to abuse of performance-enhancing drugs. A noteworthy exception to this rule is when an individual's use of psychostimulant medications to improve academic performance or assist with weight loss develops into psychostimulant abuse or dependence. Instead, most addiction specialists have only heard about the numbers of people abusing steroids or growth hormone and competing in the Olympics; very few of us have actually assessed or treated athletes who failed performance-enhancing drug tests. This diagnosis tends to be under-reported among professional athletes as well as high school, collegiate, and weekend competitors. As a result, drug abuse programs and treatment experts are rarely sought out to assist a patient who has won a race by "cheating" (i.e., taking a drug or medication that enhanced their performance). Rather, addiction treatment programs get 95% or more of their referrals from family, friends, the courts, and others concerned with the patient's addiction.

On the other hand, addiction treatment programs do evaluate patients who are referred and evaluated for addictive disease after inappropriate treatment for a psychiatric disease or symptom. For example, a patient may present to his primary care provider complaining of depression and, after a cursory evaluation, be given antidepressants. Only upon the patient's subsequent admission to a drug treatment program does it become clear that the patient was a cocaine addict. Some have called these misdiagnoses a part of the practice of "cosmetic psychiatry." But, since physicians (including psychiatrists) do not typically screen their patients for alcohol or other substance abuse or dependence beyond the occasional general question, this appears to be developing into an "Emperor's Clothes" phenomenon. By rarely ordering testing for substance abuse, physicians have set themselves apart from most other experts. For example, law enforcement officers provide the most common source of professional referral to addiction programs, due to their utilization of behavioral and drug tests for

[Haworth co-indexing entry note]: "Introduction." Gold, Mark S., and Lisa J. Merlo. Co-published simultaneously in *Journal of Addictive Diseases* (The Haworth Medical Press, an imprint of The Haworth Press, Inc.) Vol. 25, Supplement No. 1, 2007, pp. 1-4; and: *Performance-Enhancing Medications and Drugs of Abuse* (ed: Mark S. Gold) The Haworth Medical Press, an imprint of The Haworth Press, Inc., 2007, pp. 1-4. Single or multiple copies of this article are available for a fee from The Haworth Document Delivery Service [1-800-HAWORTH, 9:00 a.m. - 5:00 p.m. (EST). E-mail address: docdelivery@haworthpress.com].

impairment and intoxication. On the other hand, physicians seem to believe that they should be able to identify patients who become intoxicated, abuse substances, or suffer from drug or alcohol dependence simply by asking them. When a patient does mention drug abuse to a physician, it tends to be in reference to self-medication or a primary psychiatric problem. As a result, performance enhancement is assumed by physicians to be operational in many cases of co-morbid addiction, as the patient explains that they are taking drug 'x' to sleep and drug 'y' to lose weight and smoking drug 'z' to stay alert or "connected."

In order to facilitate the diagnostic process, Washington University's Feighner criteria and then the Research Diagnostic Criteria were developed to provide psychiatrists with agreed-upon sets of patient behaviors that were believed to comprise various diagnostic entities. As expected, the criteria were not perfect. Clinicians and researchers argued about the relevance of the proposed substance intoxication, abuse, and addiction diagnoses. Of particular note, professionals debated the logic of allowing physicians to make a drug or alcohol intoxication diagnosis without confirmatory drug testing. Later, test/retest reliability was improved, as mental disorders were routinely diagnosed by psychiatrists, other physicians, and health providers using a checklist/menu-style scheme now called the Diagnostic and Statistical Manual of Mental Disorders, fourth edition (DSM-IV). The DSM has become the standard for diagnosis in the United States and internationally. Yet, it too is not without significant problems. Based on this type of diagnostic system, it could be concluded that the United States of America is the most mentally ill country in the history of the world. In fact, the DSM criteria have recently been criticized by experts outside of the United States for generating a system that diagnoses most Americans with a psychiatric illness. The current diagnostic scheme appears to serve a self-fulfilling prophecy and is disconnected from pathophysiology, biochemical markers, and even response to treatment. For example, if SSRI response confirmed a diagnosis, then anxiety and depressive disorders would be the same diagnosis. In addition, as these checklist diagnoses became mainstream, diagnosis and diagnosis-related publishing be-

came a major enterprise for the American Psychiatric Association. Standardized, though far-from-ideal, diagnoses were then embraced by third party payers and the pharmaceutical industry. Now with the DSM in revision, pharmaceutical "stakeholders" have sought to influence Committee members' opinions regarding the diagnoses in ways that would promote the use of their medications. Unfortunately, the diagnostic system is broken and the experts assigned to fix it have to consider not only their science-based point of view, but also their careers, grants, and in some cases, retainers, industry support, and consulting agreements before actively participating in the necessary radical and important changes.

Without biological or independent markers of the disease state, independently confirming a diagnosis has been impossible. In addition, without a connection to pathophysiology, a psychiatric diagnosis cannot predict appropriate treatment. For example, anti-convulsants, SSRIs, and even antipsychotic medications can be effective in a variety of psychiatric illnesses. Indeed, the late Robert Byck, MD, suggested that SSRIs should be the first step in the diagnostic decision tree. As a result, addiction treatment center specialists often believe that psychiatric diagnoses are arbitrary and do not reflect a disease state. Misdiagnosis of the recovering patient is a common part of the addict's life story and Alcoholics Anonymous (AA) meetings are filled with psychiatrist jokes and stories. Yet, incredibly, 26.2 percent of Americans ages 18 and older–about one in four adults–suffer from a diagnosable mental disorder in any given year.[1] When applied to the 2004 U.S. Census residential population estimate for ages 18 and older, this figure translates to an amazing 57.7 million mentally ill people in the United States.

With so many in need of psychiatry and medications, it has been easy for physicians to prescribe SSRIs, antipsychotics, and other medications to many people who do not have an actual psychiatric problem. This began the current trend of cosmetic psychiatric evaluation/ treatment, which has led to the newest phase: patient self-diagnosis and symptom-specific self-treatment. Modern medicine and psychiatry have produced practitioners who advocate the widespread use of cosmetic psychiatric treatments such as taking Ritalin to improve studying for exams and SAT performance, or to reverse the effects of cannabis on college studies, or taking SSRIs for PMS or introversion. It is no surprise that physicians frequently over-prescribe psychotropic medications. The diagnostic system supports such generosity. At a major state university, the large number of students asking for psychostimulants at exam time or to reverse the cognitive effects of chronic cannabis use suggests the change from patient to client and from physician to customer-friendly agent. The result is that over half of the U.S. population now has a psychiatric diagnosis, and prescription medication misuse has become the fastest growing drug abuse trend. A recent cover story in the *Harvard Magazine* detailed the practitioner side of this issue while the *New York Magazine's* "What Pill Are You On" cover story reported the patient side of prescription sharing and self medication.[2,3]

"Performance enhancement user" is a term reserved for the person who identifies some aspect of his personality or behavior that would be improved by taking a medication or drug. These individuals seek out a willing physician (if it is a prescription they desire), or obtain it from a friend or relative. Even if the individual seeks a drug of abuse available on the street, s/he is likely to try to acquire it again from a relative or friend. Prescriptions-on-demand or prescription fee-for-service in the era of the 12-minute office visit is not far from everyday reality. *Harvard Magazine* captures the essence of this new psychiatry. Patients arrive to office visits prepared to ask for a specific prescription as a result of TV advertising, peer infomercials, internet searching, and the prevailing anti-diagnosis 'drug-for-symptom' culture. Physicians express concern that the patient may not return if they do not collaborate in these "treatments."

While many prescription and illicit drug users report symptom relief or performance enhancement, it is important to separate acute from chronic drug effects. We know that use of prescription opioids may be perceived by abusers as acutely performance-enhancing, but use does not enhance performance. For example, among anesthesiologist addicts (a group we have recently been studying), data show that the acquisition of a chronic lifelong relapsing illness is only one of the risks of such "perfor-

mance enhancement."[4,5,6] Anesthesiologists have higher rates of suicide and drug-related death, and compared to other physicians, the highest rates of all-cause mortality. One study showed that anesthesiologists have the lowest mean age at death; 64.1 years vs. 72.9 years for all doctors. Their mortality rate was shown to be 46 percent higher than the mean for all physicians.[7] In general, we prefer to assume that all use is dangerous until proven safe, regardless of the rationalization for use.

In this publication, we included papers that present and review new data on the use of medications and drugs of abuse for performance enhancement. While steroids and hGH for sports performance enhancement are frequently covered in the headlines and are the drugs and prescriptions most experts associate with performance enhancement, they are just the most visible. The prevalence of performance-enhancing use of steroids, hGH, psychostimulant medications, erectile dysfunction medications, over-the-counter nicotine-replacement products, and other illicit drugs is not well studied. Yet, U.S. household residents are more likely to report nonmedical use of prescription medications than the use of almost all illicit drugs according to the 2005 National Survey on Drug Use and Health.[8] One in twenty persons reported using prescription pain relievers nonmedically in the past year, and sizeable numbers report nonmedical use of prescription tranquilizers and stimulants. This new market for medications and drugs is important today, but these drugs may be an even more important future public health issue, as "performance-enhancing" use may become a new pathway to addiction. For these reasons, patients should be followed longitudinally to determine who and how many become a new cohort of abusers and addicts.

Mark S. Gold, MD
Chief of Addiction Medicine Division
Departments of Psychiatry, Neuroscience,
Anesthesiology, and Community Health &
Family Medicine
University of Florida College of Medicine

Lisa J. Merlo, PhD
Department of Psychiatry
McKnight Brain Institute
University of Florida College of Medicine

REFERENCES

1. Kessler RC, Chiu WT, Demler O, Walters EE. Prevalence, severity, and comorbidity of twelve-month DSM-IV disorders in the National Comorbidity Survey Replication (NCS-R). Archives of General Psychiatry. 2005; 62(6):617-27.

2. Pettus A. "Psychiatry by Prescription." Harvard Magazine. July-Aug 2006; 38+.

3. Levy A. "Pill Culture Pops." New York Magazine. 9 June 2003.

4. Gold MS, Byars JA, Frost-Pineda K. Occupational exposure and addictions for physicians: case studies and theoretical implications. Psychiatr Clin North Am. 2004; 27(4): 745-53.

5. McAuliffe PF, Gold MS, Bajpai L, Merves ML, Frost-Pineda K, Pomm RM, Goldberger BA, Melker RJ, Cendan JC. Second-hand exposure to aerosolized intravenous anesthetics propofol and fentanyl may cause sensitization and subsequent opiate addiction among anesthesiologists and surgeons. Med Hypotheses. 2006; 66(5):874-82.

6. Gold MS, Melker RJ, Dennis DM, Morey TE, Bajpai LK, Pomm R, Frost-Pineda K. Fentanyl abuse and dependence: further evidence for second hand exposure hypothesis. J Addict Dis. 2006; 25(1):15-21.

7. Svardsudd K, Wedel H, Gordh T Jr. Mortality rates among Swedish physicians: a population-based nationwide study with special reference to anesthesiologists. Acta Anasthesiol Scand. 2002; 46(10):1187-95.

8. Substance Abuse and Mental Health Services Administration. Results from the 2005 National Survey on Drug Use and Health: National Findings (Office of Applied Studies, NSDUH Series H-30, DHHS Publication No. SMA 06-4194). Rockville, MD. 2006. Available at *www.oas.samhsa.gov/nsduh/2k5nsduh/2k5Results.pdf.*

doi:10.1300/J069v25S01_01

A Review of Performance-Enhancing Drugs in Professional Sports and Their Spread to Amateur Athletics, Adolescents, and Other At-Risk Populations

David M. Martin, PhD
David A. Baron, DO
Mark S. Gold, MD

SUMMARY. Over the past 30 years, we have been involved in the establishment of a number of national and international sports drug testing programs, the review of positive tests, and the medical treatment of substance-abusing athletes. It was expected that with educational programs, testing, and supportive medical treatment this substance-abusing behavior that could lead to deadly addictive disorders would decrease. Unfortunately, this has not been the case. In fact, new, more powerful, and undetectable performance-enhancing drugs are now abused by professional athletes, and sophisticated networks of distribution have developed. Professional athletes are often the role models of adolescent and young adult populations who at times mimic their behaviors, even the abuse of drugs. This review of performance enhancement in sports is to inform addiction treatment professionals of the historical basis of performance-enhancing drugs, its institutional nature, and its spread to vulnerable athletic and non-athletic populations. doi:10.1300/J069v25S01_02 *[Article copies available for a fee from The Haworth Document Delivery Service: 1-800-HAWORTH. E-mail address: <docdelivery@haworthpress.com> Website: <http://www.HaworthPress.com> © 2007 by The Haworth Press, Inc. All rights reserved.]*

KEYWORDS. Doping, sport, athlete, hormone, steroids, EPO, hGH, adolescents, performance enhancement

David M. Martin is affiliated with JMJ Technologies, Inc., 1785 Allentown Road #185, Lansdale, PA 19446.

David A. Baron is Chairman, Department of Psychiatry, Temple University College of Medicine, Episcopal Campus, 100 East Lehigh Avenue, Philadelphia, PA 19125.

Mark S. Gold is affiliated with the Departments of Psychiatry, Neuroscience, Anesthesiology, Community Health and Family Medicine, Division of Addictive Medicine, University of Florida College of Medicine, P.O. Box 100183, Gainesville, FL 32610.

Address correspondence to: Dr. David A. Baron, Chairman and Psychiatrist in Chief, Department of Psychiatry, Temple University College of Medicine, Episcopal Campus, 100 East Lehigh Avenue, Philadelphia, PA 19125 (E-mail: dbaron@temple.edu).

The authors wish to acknowledge the invaluable assistance provided by Marita J. Krivda, MS, Director of Library Services, Temple University Medical Library without whose help this manuscript could not have been prepared.

[Haworth co-indexing entry note]: "A Review of Performance-Enhancing Drugs in Professional Sports and Their Spread to Amateur Athletics, Adolescents, and Other At-Risk Populations." Martin, David M., David A. Baron, and Mark S. Gold. Co-published simultaneously in *Journal of Addictive Diseases* (The Haworth Medical Press, an imprint of The Haworth Press, Inc.) Vol. 25, Supplement No. 1, 2007, pp. 5-15; and: *Performance-Enhancing Medications and Drugs of Abuse* (ed: Mark S. Gold) The Haworth Medical Press, an imprint of The Haworth Press, Inc., 2007, pp. 5-15. Single or multiple copies of this article are available for a fee from The Haworth Document Delivery Service [1-800-HAWORTH, 9:00 a.m. - 5:00 p.m. (EST). E-mail address: docdelivery@haworthpress.com].

INTRODUCTION

The Creed of the Olympics states, "The important thing in the Games is not winning but taking part. The essential thing is not conquering, but fighting well." As noble a goal as this is, it has little to do with the reality of the modern sports world. Athletes are rewarded by winning at virtually every level of competition. Second place is viewed as the "first loser." A coach's job security is directly related to his team's success, not that they are simply "fighting well." Given this reality it is not surprising that athletes and coaches will sacrifice and risk a great deal in order to obtain a competitive edge and enhance performance at all costs. Performance enhancement in Olympic and Professional sport has now become a medical, ethical, and legal problem for modern athletes and athletic organizations. This is primarily due to the amount of money associated with winning in today's sports industry. Multimillion dollar contracts, appearance fees, international endorsement and sports merchandising is a billion dollar industry that offers today's athlete, their sponsors and entourage previously unheard of financial gains.

Athletic performance enhancement can be gained using various diets, training routines and hard work. However, it can and has been achieved since ancient competitions by using a wide variety of physiological, mechanical and pharmacological techniques. As prize money and endorsement rewards increased, so did the science and abuse of performance-enhancing techniques. Today no sport is spared the cloud of cheating by using illegal performance enhancement. Driven by the millions of dollars now routinely available for winning a sporting event, unethical pharmacists, medical professionals, trainers and sports organizations have worked secretly, and at times without their athlete's consent, to develop sophisticated drug programs where performance is optimized, often at the risk of the athlete's health. Now, these same performance-enhancing drugs are moving out of the professional sports market to our youth and other at-risk populations at alarming rates. There are several hundred forms of known and potentially more unknown performance-enhancing drugs abused by professional athletes worldwide. This review will provide a summary of the history of performance enhancement drugs and focus on the most prevalent abused hormones; anabolic-androgenic steroids, human growth hormone (hGH) and erythropoietin (EPO). Anabolic-androgenic steroids are still the most commonly abused performance-enhancing drug while hGH and EPO are relatively new and emerging as significant hormones in athletic and other at-risk populations.

HISTORICAL OVERVIEW OF PERFORMANCE-ENHANCING DRUGS

Performance-enhancing drugs are not unique to modern athletic competition. Mushrooms, plants, and mixtures of wine and herbs were used by ancient Greek Olympic athletes and Roman Gladiators competing in Circus Maximus dating back to 776 BC. Various plants were used for their stimulant effects in speed and endurance events as well as to mask pain, allowing injured athletes to continue competing.[1-3] Since athletic competition began; adolescent males and females were put into the hands of coaches who were to mentor and guide them to perform at the highest level possible. Unfortunately, the long-term health of these athletes was not often the coaches' top priority. Often coaches were entrepreneurs that used sport and athletes to advertise and market their products. This came into clear focus in 1886 when an English cyclist, Arthur Linton, died during competition on an overdose of a mixture of ether and strychnine given to him by his coach, who was also the owner of a bicycle factory and sold equipment to competitors.[4] This was the first recorded death in sport related to performance-enhancing drugs or "doping." The origin of the word doping is said to go back to a Kafferndialekt dialect spoken in southeast Africa. The term "dope" referred to a high percent alcohol liquor and herb mixture which was used by the natives with their rituals and ceremonies.[5] This term was picked up by the English army and "doping" was first found in an English dictionary in 1888 as "a mixture of opium and alcohol given to race horses to increase speed." Performance enhancement in sport is internationally referred to as "doping" and programs to stop this activity are referred to as "anti-doping programs."

In 1904 Olympics marathon runner, Thomas Hicks, was using a mixture of brandy and strychnine and nearly died. Mixtures of strychnine, heroin, cocaine, and caffeine were used widely by athletes and each coach or team developed its own unique secret formulae. This was common practice until heroin and cocaine became available only by prescription in the 1920s. During the 1930s, it was amphetamines that replaced strychnine as the stimulant of choice for athletes. Amphetamines and male hormones were used by the Nazis in World War II to make their storm troopers more aggressive and stronger.[6] In the 1950s, the Soviet Olympic team first used male hormones to increase strength and power. When the Berlin Wall fell, the East German government's program of performance enhancement by meticulous administration of steroids and other drugs to young athletes was exposed. These well-documented and controlled hormonal-doping experiments on adolescent athletes by the East German Sports Medical Service (SMD) yielded a crop of gold medalists: mostly young females as they responded more dramatically to androgens. These athletes suffered severe medical abnormalities, including premature death. These secret doping experiments were mainly conducted at the Research Institute for Physical Culture and Sports in Leipzig, and at the Central Doping Control Laboratory in Kreischa.[7,8]

The world became acutely aware of the extent and benefits of doping in sport when Ben Johnson's Gold Medal was stripped in the 1988 Seoul Olympics for using the steroid stanazalol.[9] The IOC Medical Commission had established a list of prohibited substances in 1967 and first established anti-doping testing of athletes in the 1972 Munich Games. Effective and reliable drug testing was now in place by the Seoul Games, or at least the sports viewing public thought so. It was clear at this point that doping did work and, if gone undetected, would win Gold Medals. East German scientists from the State-run doping programs at Kreischa and Leipzig, who were disgraced in their own countries, where now in demand in Asia, former Soviet Block nations and sports organizations worldwide that wanted to promote their status in the sports world by winning Olympic Gold.

Doping became so prevalent in Olympic sport that some argued that all records should be discarded or put on hold until all forms of doping could be detected and stopped. Unfortunately, to this day anti-doping tests cannot identify all performance-enhancing drugs. It is clear that doping provides performance enhancement at considerable health risks to athletes, yet the financial rewards in modern sport are so great the consequences are eclipsed. Success in sporting events is now measured in one thousandths of a second. If doping could help improve performance, by even a thousandth of a second, it may be perceived as worth the risk by the trainer, sports organization, and athlete. Through the 1980s and 1990s clandestine doping programs spread from sport to sport guided by modern, albeit unethical, pharmacists and sports medicine professionals. In 1999, the IOC organized a World Conference on Doping in Sport in response to a shocking discovery of massive amounts of performance enhancing drugs and paraphernalia by French Police at the 1998 Tour de France. It was at this meeting that an independent global agency was founded, the World Anti-Doping Agency or WADA. Its mission was to work independently of the IOC, sports organizations and governments to lead the fight against doping in sport.[10]

Despite aggressive drug testing by the USOC and NCAA and in some professional sports, steroid abuse scandals involving high profile athletes continue to be front page news across the country. Professional sports were not subject to Olympic testing programs yet they recognized the dangers and negative consequences of drug abuse to the athlete's health and to the sport's image. Anti-doping programs were put into place in professional Football, but Baseball and Hockey players' unions and collective bargaining agreements prevented comprehensive testing to be put into place. Then in 1998 when Mark McGuire broke Roger Marris' home run record, it was revealed that he had been taking a supplement containing a precursor to nandrolone, a steroid. At that time Major League Baseball did not ban steroids and did not believe that steroids were a problem within the league. However, subsequent congressional hearings and former players revealed steroid use was a problem in Major League Baseball.[11,12] The short- and long-term medical effects of doping that led to headline articles such as "The Bonds Bombshell" and, "Tainted by Drugs" are yet fully understood and an important area for ongoing investigation by sports medicine professionals.[13] However, it is clear that serious liver, kidney, endocrine, cardiac, and psychiatric side effects can result from steroid use.[14]

In June of 2003, another significant event occurred. A syringe was sent to a WADA-accredited laboratory in Los Angeles that contained a "designer" steroid that was not known and not on the current IOC prohibited list. It contained tetrahydrogestrinone or THG, a synthetic steroid made specifically to avoid detection by modern anti-doping technologies.[15] This led to a series of investigations resulting in the indictment and subsequent conviction of individuals running a performance-enhancing program for professional athletes at the BALCO pharmacy in San Fransico.[16] In May of 2006 Spanish police arrested five people and seized a variety of banned performance-enhancing drugs and blood-doping supplies at a Madrid doping clinic. Here, professional athletes (mostly cyclists) would receive medically-supervised injections of hormones and other performance-enhancing drug regimes.[17] The 40-page police report included a clear paper trail of doping procedures on at least 50 professional cyclists. The report was given to the International Cycling Union which led to the disqualification of 23 professional cyclists; virtually all the top contenders from the 2006 Tour de France. Documents from the Madrid doping clinic also implicated other sports as being doping clients in addition to professional cyclists. In a separate investigation that same week in France resulted in 23 individuals sentenced to 4 years in jail for trafficking a cocktail of amphetamines and other performance-enhancing drugs known in the cycling world as "Belgium Pot" to professional cyclists.[18] In the June 2006 issue of the *Journal of Applied Physiology*, an article from Stanford University reported that Viagra can be used to increase the performance of cyclists in high altitudes by approximately 45%, suggesting a whole new class of performance-enhancing drugs not restricted to cycling.[19] This historical overview suggests not only the continual abuse of performance-enhancing drugs by athletes but also the international development of a clandestine and sophisticated distribution network for performance-enhancing drugs that follows the modern sports industry. Today performance-enhancing programs and drugs are not the exclusive province of elite athletes but have spread to health clubs, high schools and other at-risk populations creating a $1.4 Billion dollar industry that is growing daily as new compounds are synthesized and marketed.[20]

POPULAR PERFORMANCE-ENHANCING DRUGS

There are literally hundreds of known performance-enhancing drugs and an equal number of designer, veterinary, and yet to be identified drugs abused in sports today. The 2006 World Anti-Doping Agency list of prohibited substances include the following major categories: *S1 Anabolic Agents* include exogenous anabolic androgenic steroids (androstendiol, boldenose, closterbol, danazol, etc.), endogenous anabolic androgenic steroids (dihydroxytestosterone, testosterone and other naturally occurring steroids) and other anabolic agents (clenbuterol, tibolone etc.).[21] In all there are over a hundred natural and synthetic steroids, precursors and metabolites that are banned by WADA. *S2 Hormones and Related Substances* include Erythropoietin (EPO), Growth Hormone (hGH), Insulin-like Growth Factors (IGF-1) Mechno Growth Factors (MGF), Gonadotropins (LH, hCG for males only), Insulin and Corticotrophins. *S3 prohibits Beta-2 Agonists* (terbutaline, salbutamol, etc.), *S4 prohibits Agents with Anti-Estrogenic Activity* (anastrozole, letrozole, clomiphene etc.), and *S5 prohibits Diuretics and other Masking Agents* (furosemeide, hydrochlorothiazide, etc., masking agents such as epitestosterone, probenicid, plasma expanders, etc.). *S6 includes Stimulants* (amphetamines, ephedrine cocaine, etc.), *S7 Narcotics* (morphine, oxycodone, etc.), *S8 Cannabinoids* (marijuana, hashish) and *S9 Glucocorticoseroids* (allowed externally but not but not internally). WADA also lists prohibited methods which include: *M1 Enhancement of Oxygen Transfer* (blood doping, efaproxial, etc.), *M2 Chemical and Physical Manipulation* (tampering or substitution of sample) and *M3 Gene Doping*. In addition,

WADA prohibits *P1 Alcohol* (in-competition and specific sports archery, billiard, etc.) and *P2 Beta-Blockers* (specific sports archery, billiard, etc.).

Testing for the above list of compounds is technically challenging, expensive (several hundred dollars) and only performed by 35 WADA accredited laboratories worldwide. There is only one WADA accredited lab in the United States located at UCLA that was established for the 1984 Olympic Games in Los Angeles.[22] This lab recently identified the designer steroid tetrahydrogestrinone (THG).[23] Because athletes who are doping would like to have their urines screened prior to competing, WADA laboratories only perform tests for verified professional and Olympic sports organizations. This has not discouraged doping as clandestine testing of athletes and supply of drugs is done worldwide by black market laboratories and pharmacies. As the athletes, their supporters and suppliers become more sophisticated in their doping programs, this information became available on the internet and performance-enhancing drugs spread into our high schools, colleges and gyms nationwide.[24] The abuse of performance-enhancing drugs in adolescents was first reported in the 1980s, focusing on steroids. This review restates this continuing problem in light of the ease of availability of these drugs, increased trafficking worldwide, and Hollywood's message that youth, success and fortune, often represented in sports icons, are culturally desirable even at the cost of ethics, personal health, and safety. Steroids are still the most detected performance enhancement drug by WADA laboratories. However, because of the limitations of laboratory technology and sophistication of athletes are doping to avoid detection; they may not be the most abused. Today human growth hormone (hGH) which builds muscle, and erythropoietin (EPO) which increases oxygen transport, have emerged as doping problems as well and are used in combination with steroids.[25] Because of the widespread abuse and availability of these three performance-enhancing hormones, this review will focus on these three agents as examples of the continuing problem of drug abuse in professional, adolescent, and weekend athletes.

ANABOLIC ANDROGENIC STEROIDS

Anabolic androgenic steroids are naturally occurring male hormones involved in a wide range of physiological functions. Testosterone is the most commonly known steroid, the term anabolic means "building" and androgenic relates to male characteristics. Simply referred to as "steroids" they fall into two categories: endogenous or naturally occurring like testosterone and exogenous or synthetic like danazol.

In 1923 Bob Hoffman, a World War I veteran, invented the barbell and formed the famous York Barbell Company. A dominant figure in US Weightlifting, he published "Strength and Health" magazine and sold health and food supplements in his gym. As a weightlifting coach his success led to him being named the head coach of the US Olympic Weightlifting team and he became friends with the team physician, John Ziegler, MD. At the 1954 World Championships in Vienna, Bob Hoffman met with a Soviet colleague who told him of a synthetic form of testosterone developed by the Nazi's which produced dramatic improvements in strength and power. Dr. Ziegler contacted CIBA Pharmaceuticals in pursuit of synthetic testosterone. CIBA had conducted a number of studies on the use of synthetic testosterone in pain patients and the physically disabled. Dr. Ziegler decided to study its effects in weightlifting athletes which resulted in the development and first mass produced anabolic steroid, danazol.[26,27]

First abused in Olympic sports in the 1950s, the abuse of steroids in young male athletes who were not in Olympic training began being reported in the 1980s.[28-30] As demand increased, trafficking steroids at schools and gyms become common and the use of steroids was seen in younger and younger populations.[31] Steroid sources included doctors, trainers, friends, the black market and foreign suppliers. The Anabolic Steroid Enforcement Act of 1990 placed certain anabolic steroids in Schedule III of the Controlled Substances Act (CSA). Previously, steroids had been unscheduled and controlled only by state and local laws. The Anabolic Steroid Enforcement Act brought anabolic steroids under the record-keeping, reporting, security, prescribing, import and controls of the CSA. Because steroids were now classified as a Schedule III substance, all manufacturers and distributors of steroids were required to register with the Drug Enforcement Agency (DEA). However, the amount of illegal steroids entering the United States has increased dramatically, forming an estimated 100 million dollar black market for steroids with greater than 80% manufactured in Mexico.[32]

The serious medical side effects of steroids described in the medical literature include liver function abnormalities, liver and kidney tumors, endocrine and reproductive dysfunctions, testicular atrophy, lipid and cardiac effects and psychiatric symptoms.[33] These consequences are also exaggerated with the common doping practices using ten times or more the recommended medical dose and repeated use of multiple drugs or 'stacking,' e.g., steroids and EPO or hGH. Added to this, a new problem has emerged with the manufacture of "counterfeit" drugs by unregulated pharmacies that often produce drugs that are tainted with impurities, contain no medication, or are potentially harmful.[34] This problem of counterfeit drugs is not restricted to anabolic steroids, many performance-enhancing drugs and ethical medications are now being counterfeited and introduced into the pharmacy market. These medications look and are packaged exactly as the manufacturer's finished product intended for ethical medical use. Now, more so than in the past, when an athlete buys performance-enhancing drugs from a friend or at the gym he will never know exactly what is being bought or taken. Steroids are sold on the internet ranging in price range from $50-$200 per regime depending upon the type of steroid and doping program selected. These black market steroids may or may not contain any medication at all or may contain harmful material. Testing for steroids in urine is available at a few commercial clinical laboratories in the United States and can be obtained in the price range of $100-$200/test depending upon the number of steroids screened. WADA-certified labs only deal with official sports organizations and sports program administrators.

HUMAN GROWTH HORMONE (hGH AND rhGH)

hGH is a naturally occurring polypeptide hormone produced by the anterior pituitary gland and is one of the major hormones influencing growth and development in humans. The major function of hGH is protein synthesis through fat metabolism, speeding muscle tissue repair, building and tone. This has created a demand for hGH not only in sports but in the new and controversial "anti-aging" medicine market. It also is involved in the control of normal linear growth (height) from birth to adulthood as well as the regulation of energy production and storage. Dr. Harvey Cushing first discovered hGH in 1912 and isolated it from human and monkey cadaver brains by glandular extraction of the pituitary in 1956. Two years later it was used to treat dwarfism in children by injection. The unfortunate development of Creutzfeldt-Jakob disease, a degenerative brain disorder, in boys who were treated with cadaver growth hormone led to the discontinuation of all products derived from the human pituitary gland.[35] Because of this ban the abuse of hGH was rare in sport until the middle to the end of the 1980s. In 1985 Genentech received approval from the U.S. Food and Drug Administration (FDA) to market Protropin® (somatrem for injection) growth hormone for children with growth hormone deficiency. This was the first recombinant DNA form of growth hormone (rhGH) that was safer than cadaver extracts used in the past. Recombinant DNA technology made the production of pharmaceutical grade growth hormone easier and cheaper. Genetically engineered rhGH is available and is marketed as Nutropin (*Genentech*), Humatrope (*Lilly*), Genotropin (*Pfizer*), Norditropin (*Novo*), Saizen (*Serono*), and Tev-Tropoin (*Teva*). Most human growth hormone used in medicine and diverted to sports doping is now obtained by recombinant technology, and is simply referred to as hGH (but it may also appear as rhGH or HGH). Unfortunately, cadaver extracts of pituitary hGH may still be in circulation. It has been reported that a Russian coach was arrested and upon searching his apartment in Moscow, over 1000 cadaver pituitary glands were found preserved in a large container.[36] All forms of pharmaceutical grade hGH are reported to have equal potency and effects and are primarily used to treat dwarfism, endocrine disorders, major tissue surgery and wasting diseases. However, the same problem of counterfeit drugs exists with trafficking hGH–illegal pharmaceutical manufacturers are now flooding the black market with hGH vials of unknown quality and safety. It is estimated that an eight-week performance enhancement regime of pharmaceutical grade rhGH will cost about $2000, well out of the range of an adolescent and the majority of weekend athletes. However, the increased trafficking of low cost counterfeit rhGH will create interest and experimentation in these at-risk populations. hGH is marketed on the internet in many forms: pills, drops and aerosol formulations; most are ineffective and shams. The normal route of administration of hGH is injection, posing an additional health risk of infection from non-sterile counterfeit drugs and the risk of HIV and hepatitis transmission caused by shared needles.

Olympic, professional, and weekend athletes abuse hGH because of unsubstantiated reports that it is as effective as anabolic steroids with fewer side effects. They use hGH as a steroid substitute to prevent loss of muscle after discontinuing the use of steroids. Ben Johnson admitted to using hGH along with steroids during investigations after his disqualification in Seoul. According to some controlled scientific studies, hGH does not increase muscle strength, however, most reports on hGH performance enhancement in sport are limited in scope, anecdotal or based on animal studies.[37] Nevertheless, the abuse of hGH in sports is escalating with large caches of needles and vials of hGH being confiscated at sporting events worldwide. Six months prior to the 2000 Summer Olympic Games, a pharmacy in Sydney was broken into and 1,575 multiple dose vials of hGH were taken while nothing else was touched. Also, on their way to Australia, the Chinese swimming team was detained as needles, syringes, and vials of hGH were found by Customs officials in their baggage.

It was not until the 2004 Summer Olympic Games in Athens did WADA-certified labs have a test for hGH abuse, yet testing at the Games did not result in any positives for hGH. This may be in part due to the fact that the technology was new and WADA laboratories had little experience with the test. The side effects of using hGH may lead to life-threatening health conditions, especially since some estimates report that athletes who use hGH to enhance performance are taking 10 times the therapeutic dosage. Some reported side effects of hGH are abnormal bone growth, hypertension, cardiovascular disease, cardiomyopathy, glucose intolerance, colonic polyps, decreased life span, and cancer.[38]

Since hGH is a naturally-produced hormone and rhGH is similar in structure, testing for doping with rhGH has been a technical challenge only recently solved by WADA-certified laboratories. Routine blood tests for hGH available at clinical laboratories will not differentiate hGH from rhGH and are of no value in determining if an adolescent or weekend athlete is doping. Testing for steroids still may be helpful as hGH is often used in combination, stacked or after a course of steroids. The threat of testing alone is a powerful deterrent to substance abuse and helpful in eliciting drug abuse disclosures when obtaining a medical history.

ERYTHROPOIETIN (EPO)

Erythropoiefin (EPO) is a naturally occurring hormone produced by the kidney that stimulates red blood cell production in the bone marrow in response to low circulating oxygen levels in humans. Like hGH it was genetically engineered in 1985 and is commonly used to treat anemia caused by renal failure, in chemotherapy, and in AZT treatment of AIDS. The history of EPO is not as extensive as hGH as it was only believed to exist for most of the twentieth century. It was not until the 1960s when its source was identified as the kidneys, and not until 1977 was EPO first identified and extracted from human urine. This was concurrent with the development of recombinant DNA technology and in 1989 Epogen (Amgen) was released in the United States and approved for the treatment of anemia. Procrit (J&J) was licensed from Amgen in 1991 for the treatment of chemotherapy-induced anemia. European formulations include Aranesp (Amgen), Eprex (J&J), and NeoRecorman (Roche).

EPO abuse in sport was believed to start as soon as the drug was available as a replacement for the older, more complex and dangerous doping technique referred to as "blood doping." In this technique an athlete donates his own blood several months before a competition, stores it and transfuses it back into himself prior to competing. This technique is fraught with problems and health risk. EPO accomplishes this same effect by increasing red blood cells which results in more oxygen in circulation. It was in 1998 at the Tour de France that French Customs arrested Willy Voet, a physiotherapist of the Festina Cycling Team, for the illegal possession of needles, syringes and over 400 bottles containing EPO, hGH, steroids, amphetamines, narcotics and stimulants. In the following weeks French police raided several hotel rooms of cycling teams and continued to confiscate illegal doping agents. In the criminal trials that followed, it was revealed that sophisticated doping programs were common and encouraged in cycling as well as other sports.[39]

EPO used for medical treatments can cost thousands of dollars a month and is administered by intravenous or subcutaneous injection. As with steroids and hGH, doping with EPO is often injected in supernormal doses that could cause increased blood viscosity, deep vein thromboses, pulmonary emboli, coronary thromboses, cerebral thromboses, pulmonary embolism, arrhythmias, stroke and death. It has been estimated that 20 European cyclists have died since 1987 due to abuse of EPO, making it one of the most deadly doping agents. The genetically engineered form of EPO is indistinguishable from naturally occurring EPO, making routine blood testing useless to determine if an athlete is doping. At the 2000 Summer Games in Sydney the Australian WADA-certified laboratory first launched a sophisticated anti-doping test for EPO that required both urine and a blood sample.[40] Over 300 tests were performed for EPO for the first time in Olympic history and no positives were reported. This could be due to the fact that the technology for the test was new and questions still existed about the assay.

ADOLESCENT AND AT-RISK POPULATIONS PERFORMANCE-ENHANCEMENT DRUG ABUSE

Given the above history and current state of knowledge, it is difficult to understand why there would be over a million abusers of steroids in our nation's youth. Unlike professional athletes, they will not have fame and fortune as a result of using steroids. Perhaps the answer can begin to be found in a Sports Illustrated report of questions asked of a cohort of elite athletes in 1995. One of the questions was, "If you were given a performance enhancing substance and you would not be caught and win would you take it?" 98% of the athletes responded, "Yes." The more chilling question was, "If you were given a performance enhancing substance and you would not be caught, win all competitions for 5 years then die, would you take it?" Greater than 50% said, "Yes."[41] This is a limited and unscientific study published in a sports magazine. However, this report suggests that winning, even at the cost of death, is acceptable.

For purposes of this review, adolescent athletes are defined as those between the ages of 12 and 18 years old and not professionally competing; weekend athletes are defined as those greater than 19 years old who are engaged in regular recreational and fitness exercise but not college or amateur athletes. At-risk populations are those who are in both age groups but also engage in risky behaviors such as substance abuse, fighting, tobacco use and high risk sexual behavior. Weekend athletes are included in this report as long-term abuse of steroids can begin in adolescence. The discussion also focuses on steroids as they may be associated with potential hGH and EPO abuse. Steroids have been extensively tested for by WADA labs over the past two decades while tests for hGH and EPO have only recently been developed. Although not tested and potentially under reported, hGH and EPO are often used in combination with steroids making steroids a potential marker for the abuse of these drugs.

Pioneering studies were done by Buckley and his colleagues in the early 1980s when they interviewed 3,403 male high school seniors nationwide.[34] Their results reported in 1988 indicated that 6.6% of respondents had reported using steroids and greater than two-thirds of the group started using steroids when they were 16 years old or younger. Twenty percent reported that health professionals were the primary source for obtaining steroids and 38% used injectable steroids. Pope and colleagues studied 1,010 college men for use of steroids and also reported their findings in 1988.[29] The study found that only 2% of the respondents reported using steroids. The authors qualified their finding as potentially underestimating the true prevalence of steroid abuse. However, it is interesting to note that this study reported that 25% of those reporting using steroids were not athletes. They abused steroids to improve personal appearance, a problem that continues today and is fueled by the media and "anti-aging" marketing. A review of published reports concluded that overall between 3-12% of high school students used steroids and of the group of abusers about half were adolescent females.[42]

Contrary to popular belief and supported by Pope's early findings, steroid abuse is not exclusively related to performance enhancement. DuRant and his colleagues reported in 1993 that steroid abuse in ninth graders was associated with use of cocaine, injected drugs, alcohol abuse, marijuana, cigarettes and smokeless tobacco.[43] He then reviewed the 1991 Centers for Disease Control and Prevention Youth Risk Behavior Survey of over 12,272 male and female public and private high school students. He confirmed the earlier finding that there is an association with steroid abuse and multiple drug use.[44] In a later review of the 1997 Centers for Disease Control and Prevention Youth Risk Behavior Survey of 16,262 high school students, Miller and her group reported no significant correlation in male or female steroid-abusing high school students with physical activity, nor were athletic participation or strength conditioning alone associated with lifetime steroid abuse.[45] Rather, they found that athletic participation was less of a factor then behavior problems such as substance abuse, fighting, binge drinking, tobacco use and high risk sexual behavior. They suggest steroid abuse may be part of a much larger syndrome of problem behaviors. This is important since it expands the culturally pervasive thought that performance enhancement in sports and strength training is the primary factor for steroid abuse. In 2002, Irving and her colleagues confirmed Miller's report that physical activity was not associated with steroid abuse alone. Her group shed light on the fact that male and female adolescent steroid abuse may also be associated with unhealthy attitudes and behaviors to lose, gain or control weight and body shape.[46] Clancy and Yates reported in 1992 that steroid abusers may have a unique set of clinical differences and are distinct from other drug abusers.[47] In an attempt to characterize the adolescent steroid abuser, in 2000 Bahrke and colleagues associated a number of personal high-risk behaviors and other factors with a partially developed profile of an adolescent anabolic steroid abuser.[48] What has become evident is that not only high school and weekend athletes are potential steroid abusers, steroid abuse may also include a wider population of non-athletes who have behavioral problems and may experiment with these now easily available performance-enhancing drugs. Their motivation may not be athletic enhancement, but rather for cosmetic and body shaping purposes. To maintain youthful appearances, weekend athletes may experiment with hormones encouraged by "anti-aging" marketing, while adolescent females desirous of the long, lean female media images of "adult women" may use steroids and hGH to reduce fat and increase muscle tone.

DISCUSSION

Since the 1980s, the medical literature, news articles and anecdotal reports have repeatedly suggested there is continued abuse of performance-enhancing drugs in our adolescent populations. Access, availability and promotion of these drugs for sports and cosmetic purposes are increasing on the internet at an alarming rate. Performance-enhancing drugs are also being seen and routinely available in the booming health club market and gaining acceptance by youth worldwide.[49] Steroids, hGH, and EPO are just a few of the drugs spreading from professional sports to our youth. This has created a performance enhancement culture of acceptance and a pharmacopoeia of existing and new designer drugs available to adolescents, weekend athletes and other at-risk populations. We as a society must acknowledge and deal with the spread of performance-enhancing drugs in our youth aggressively or by our silence repeat the tragedy of institutional child abuse uncovered in the clandestine labs of Leipzig and Kreischa. Modern sports and the media's misplaced fixation on fame, fortune and winning at all costs have unintentionally created a growing market for performance-enhancing drugs. These drugs, once only abused by elite athletes are clearly spreading into our schools and health clubs nationwide. They are being accepted by a whole new generation of young customers who see reports daily in the newspapers of sports icons accused of abusing drugs only to continue playing, breaking records and claiming fortunes. These same performance-enhancing drugs are also abused by adolescents and weekend athletes and non-athletes who have wider behavioral and health risk problems. In addition, these drugs are now being abused by male and female adolescents for cosmetic purposes in an attempt to achieve the "cut" and sexy look promoted by Hollywood and the media. Continuing educational programs developed for these at-risk populations, such as the Atlas and Athena programs funded by the National Institute of Drug Abuse (NIDA) and policy statements from the Pediatrics community are important first steps to curb these dangerous behaviors.[50-52] Testing for performance-enhancing drugs in High Schools as a means of early detection, intervention and prevention is now being launched in New Jersey with other states following their lead.[53] Medical professionals, teachers, coaches and sports organizations must all be made aware of this continuing problem in our adolescent and at-risk populations and become a part of the solution by open, honest discussion. Most importantly, professional athletes and their player's associations must serve as role models and spokesmen for drug-free sport and lifestyle. This position must be actively supported by the media, owners of teams and international sports federations by providing consistent leadership and advocacy of drug-free sport, regardless of costs and consequences. This is the only way we can prevent the continuing spread and serious medical consequences of the abuse of performance-enhancing drugs in sport and its spread into our society.

REFERENCES

1. Wadler GI, Hainline B. (eds.) Drugs and the Athlete, Philadelphia, PA. FA David Co: 1989;3-4.

2. Yesalis CE. History of doping in sport. In Bahrke MS, Yesalis CE (eds). Performance Enhancing Substances in Sport and Exercise, Champaign, IL, Human Kinetics: 2002;1-20.

3. Landry GL, Kokotaio PK. Drug screening in athletic settings. Curr Probl Pediatr. 1994; 24:344-359.

4. de Merode A, Schamasch P. Harmonisation of Methods and Measures in the Fight Against Doping in Sport, Final Project Report SMT4-CT98-6530, IOC Medical Commission, Lausanne, Switzerland 1999;4-5.

5. Lippi G, Guidi G. Doping in sports. Minerva Med 1999; 90(90):345-357.

6. Ungerleider S. Faust's Gold: Inside the East German Doping Machine. St. Martin's Press, New York, NY. 2001;50:124.

7. Silverman F. Guaranteed aggression: The secret use of testosterone by Nazi troops. JAMA. May 1984; 129-131.

8. Franke WW, Berendonk B. Hormonal doping and andronogenization of athletes: A secret program of the German Democratic Republic. Clinical Chemistry. 1997; 43:1262-1279.

9. Olympic Movement Anti-Doping Code, Lausanne, Switzerland, International Olympic Committee, 1999 accessed at *http://www.olympic.org* on February 26, 2006.

10. World Anti-Doping Agency: accessed at *http://www.wada-ama.org*, on February 26, 2006.

11. Congressional Committee on Government Reform hearings, Tom Davis, Chairman, Report on Investigation into Rafael Palmeiro's March 17, 2005 Testimony before the Committee on Government Reform. November 12, 2005.

12. Conseco J. Juiced, Wild Times, Rampant 'Roids, Smash Hits and How Baseball Got Bigger. HarperCollins, New York, NY, 2005.

13. Cole C. Tainted by Drugs. National Post. December 30, 2004.

14. Hartgens F, Harm K. Effects of androgenic anabolic steroids in athletes. Sports Med. 2004; 34(8):513-554.

15. Ritter SK. Designer steroid rocks sports world, American Chemical Society, Chem Eng News. 2003; 81(46):66-69.

16. US Department of Justice, Attorney General for Northern California, Press Release, Grand Jury Returns 42-Count Indictment Charging Individuals Associated with Bay Area Lab Cooperative (BALCO), February 12, 2004.

17. Ullrich, Basso barred from Tour amid doping scandal. accessed at *www.dailyhearld.com* on July 01, 2006.

18. French court sentences 23 in doping case. Reuters. accessed at *www.washingtonpost.com* on July 3, 2006.

19. Hsu AR, Barnholt KE, Grundmann NK, Lin JH, McCallum SW, Friedlander AL. Sildenafil improves cardiac output and exercise performance during acute hypoxia, but not normoxia. J Appl Physiol. 2006; 100: 2031-2040.

20. Performance Enhancing Drugs. Healthy NJ. March 2005. accessed at *www.healthynj.org* on March 7, 2006.

21. World Anti-Doping Agency Prohibited List. accessed at *http://www.wada-ama.org/en/prohibitedlist.ch2* on February 26, 2006.

22. Catlin DH, Kammerer RC, Hatton CK, Sekera MH, Merdink JL. Analytical chemistry at the Games of the XXIIIrd Olympiad in Los Angeles, 1984. Clin Chem. 1987; 33(2 Pt 1):319-327.

23. Catlin DH, Sekera MH, Ahrens BD, Starcevic B, Chang YC, Hatton CK. Tetrahydrogestrinone: Discovery, synthesis and detection in urine. Rapid Commun Mass Spectrom. 2004;18(12):1245-1249.

24. How to buy steroids on line. accessed at *http://www.legalsteroids.com* on March 1, 2006.

25. Prendergast HM, Bannen T, Erickson TB, Honore LR. The toxic torch of the Olympic games. Vet Hum Toxiol. 2003; 45(2):97-102.

26. Goldman B. Death in the Locker Room: Steroids, Cocaine and Sports. Tucson, The Body Press, 1987.

27. When Steroids Were All the Rage. Phil. Enquirer, October 2, 2002, D1-D2.

28. Kutcher EC, Lund BC, Pery PJ. Anabolic steroids: A review for the clinician. Sports Medicine. 2002; 32:285-296.

29. Pope HG, Katz DL, Champoz R. Anabolic androgenic steroid use among 1,010 college men. Physician and Sports Medicine. 1988; 16(7): 75-81.

30. Vogel G. A race to the starting line. Science, special section on "Testing Human Limits." 2004; 305(5684): 632-635.

31. Bahrke MS, Yesalis CE, Kopstein AN, Stephens JA. Risk factors associated with anabolic-androgenic steroid use among adolescents. Sports Med. 2000; 29(6): 397-405.

32. Department of Drug Enforcement History. accessed at *http://searchjustice.usdoj.gov* on March 15, 2006.

33. Kuipers H. Anabolic steroids: Side effects. In: Encyclopedia of Sports Medicine and Science, New York. Macmillan, T. D. Fahey (Editor).1998.

34. Buckley WE, Yesalis CE 3rd, Friedl KE, Anderson WA, Streit AL, Wright JE. Estimated prevalence of anabolic steroid use among male high school seniors. JAMA. 1988; 260(23):3441-3445.

35. Brown P. Potential Epidemic of Creutzfeldt-Jakob disease for human growth hormone therapy. New Engl J Med. 1985; 313:718.

36. Sonksen PH. Insulin, growth hormone and sport. J Endocrinology. 2001; 170:13-25.

37. Jenkins PJ. Growth Hormone and Exercise: Physiology, use and abuse. Growth Hormone & IGF Research, Supp. A. 2001; 781-77.

38. Perls TT, Reisman NR, Olshansky SJ. Provision or Distribution of Growth Hormone for "Antiaging": Clinical and Legal Issues. JAMA. 2005; 294:2086-2090.

39. Voet W. Breaking the Chain: Drugs in Cycling–The True Story. Yellow Jersey Press, London, 1999.

40. Corrigan G, Kazlauskas R. Drug testing at the Sydney Olympics. MJA 2000; 173:312-313.

41. Bamberger M, Yaeger D. Over the edge. Sports Illustrated 1997; 14:62-70.

42. Lucas SE. Current perspectives on anabolic-androgenic steroid abuse. Trends in Pharm Science. 1993; 14:61-68.

43. DuRant RH, Rickert VI, Ashworth CS, Newman C, Slavens G. Use of multiple drugs among adolescents who use anabolic steroids. NEJM. 1993; 328(13):922-926.

44. DuRant RH, Escobedo LG, Heath GW. Anabolic steroid use, strength training and multiple drug use amount adolescents in the United States. Pediatrics. 1995; 96(1): 23-28.

45. Miller KE, Hoffman JH, Barnes GM, Sabo D, Melnick MJ, Farrell MP. Adolescent anabolic steroid use, gender, physical activity and other problem behaviors. Subst Use Misuse. 2005; 40:1637-1657.

46. Irving LI, Wall M, Jeumark-Sztainer D. Steroid use of adolescents: Findings for the project EAT. J Adol. Health. 2002; 30:243-252.

47. Clancy GP, Yates WR. Anabolic steroid use among substance abusers in treatment. J Clin Psychiatry. 1992; 53(3):97-100.

48. Bahrke MS, Yesalis CE. Abuse of anabolic androgenic steroids and related substances in sport and exercise. Current Opinions in Pharm. 2004; 4:614-620.

49. Laure P, Binsiger C. Adolescent athletes and the demand and supply of drugs to improve their performance. J Sport Science and Medicine. 2005; 4(3):272-277.

50. Teen Athletes and Performance Enhancing Substances: What Parents Can Do, Mayo Clinic.com, Teen's Health, December 22, 2004. accessed at *www.mayoclinic.com* on March 7, 2006.

51. The Atlas and Athena Programs, University of Oregon. accessed at *www.ohsu.edu/hpsm/index* on March 8, 2006.

52. Committee on Sports Medicine and Fitness. Policy statement on the use of performance enhancing substances. Pediatrics. 2005; 115(4):1103-1107.

53. Lawlor C. New Jersey institutes statewide steroid-testing for high school athletes. accessed at *http://www.usa.com/sports/preps/2006-06-07_steroid-testing.x.html* on November 20, 2006.

doi:10.1300/J069v25S01_02

Nicotine:
Risks and Performance-Enhancing Effects

C. Patrick Lane, BS
Noni A. Graham, MPH
Ellen A. Ovson, MD

SUMMARY. Nicotine, the active ingredient in tobacco, has been available in the United States for over 20 centuries. Nicotine and compounds effecting nicotinic acetylcholine receptor stimulation have been implicated in the improvement of a number of neuronal conditions, functions, and activities. Sixteen functional human nicotinic receptors have been identified throughout the body pre-, post-, and perisynaptically; and 12 functional types are known to exist in the CNS. Nicotine's therapeutic characteristics provide partial relief from degenerative neurologic diseases such as Alzheimer's and Parkinson's, and have been found to impact anxiety and depression, attention-deficit disorder, motor skills, cognitive functions and memory. For example, experimental animal studies have demonstrated that nicotine-induced memory improvements remain after the withdrawal of nicotine. Advances in the understanding of nicotine's action on the central nervous system (CNS) have led to insight into the pervasive and enduring nature of addiction. This paper will review studies from laboratory and clinical findings indicating the importance of CNS nicotinic mechanisms in normal human cognitive and behavioral functioning as well as their role in disease states. In addition, the efficacy of nicotine replacement therapy for smoking cessation will be reviewed in terms of demographics and response rates. doi:10.1300/J069v25S01_03 *[Article copies available for a fee from The Haworth Document Delivery Service: 1-800-HAWORTH. E-mail address: <docdelivery@ haworthpress.com> Website: <http://www.HaworthPress.com> © 2007 by The Haworth Press, Inc. All rights reserved.]*

KEYWORDS. Nicotine, CNS, nicotine replacement therapy, performance-enhancing effects

INTRODUCTION

Identification of nicotine in pipes found in archeological sites dates the use of tobacco in North America to approximately 300 BC. Although tobacco is not native to this area today, it has been assumed that its use was encouraged and it tended to be an important non-subsis-

C. Patrick Lane is a graduate student at the University of Southern Mississippi, Department of Psychology, Hattiesburg, MS.

Noni A. Graham is affiliated with the University of Florida, Division of Addiction Medicine, Gainesville, FL.

Ellen A. Ovson is a member of the American Society of Addiction Medicine, Pine Grove Behavioral Health and Addiction Services, Hattiesburg, MS.

The authors wish to thank David Echevarria, PhD, Department of Psychology, University of Southern Mississippi, Hattiesburg, MS for his enthusiastic and selfless contribution to this paper.

[Haworth co-indexing entry note]: "Nicotine: Risks and Performance-Enhancing Effects." Lane, C. Patrick, Noni A. Graham, and Ellen A. Ovson. Co-published simultaneously in *Journal of Addictive Diseases* (The Haworth Medical Press, an imprint of The Haworth Press, Inc.) Vol. 25, Supplement No. 1, 2007, pp. 17-31; and: *Performance-Enhancing Medications and Drugs of Abuse* (ed: Mark S. Gold) The Haworth Medical Press, an imprint of The Haworth Press, Inc., 2007, pp. 17-31. Single or multiple copies of this article are available for a fee from The Haworth Document Delivery Service [1-800-HAWORTH, 9:00 a.m. - 5:00 p.m. (EST). E-mail address: docdelivery@haworthpress.com].

tence agricultural product. This suggests that tobacco was important to social practices and that nicotine may have been used medicinally in North America dating back to the Historic Period.[1]

Today, nicotine has been determined to provide a number of neuroprotective and therapeutic uses, both alone and in combination with other drug therapies.[2] The intent of this paper is to describe the performance-enhancing aspects of nicotine, which include memory, attention and discrimination, and mediation of reward. A review of nicotine's ability to stimulate improved performance in certain disorders will also be considered. Finally, the literature concerning nicotine replacement therapy for smoking cessation will be considered for the potential of nicotine abuse.

Nicotine generally acts to increase brain activity through the stimulation of selective subtypes of nicotinic receptor activity and has been shown to affect a wide variety of functions ranging from gene expression, regulation of hormone secretion and enzyme activities.[3,4] It enhances visual recognition and spatial memory through activation of nicotinic acetylcholinergic pathways correlated with increased activity in neural pathways in the entorhinal cortex for improvement of memory and in pathways dedicated to vigilance.[5,6] Nicotine is well established as an enhancement to cognitive tasks, especially regarding attention and "working" or short-term memory in both animal and human models.[7,8] Nicotine use has been shown to reduce response time and to increase attention in participants diagnosed with ADHD.[7] In addition, nicotine is implicated in reducing negative affect.[9] Furthermore, nicotine administration produces highly diverse effects, encompassing changes in body temperature, locomotor activity, cardiovascular and gastrointestinal function, cortical blood flow and nociception (Table 1).

The aforementioned effects include potential therapeutic benefits for certain patient populations suffering disorders characterized by an attentional deficit and response inhibition.[7,10] Grilly[11] has found that the typical animal and human response to nicotine is similar to that of amphetamine and that enhanced performance in detection of minuscule stimulus alterations (vigilance) is distinguishable from that of clas-

TABLE 1. General effects of nicotine in organismic, cellular, and molecular levels

Organismic Effects

Increased heart rate
Increased blood pressure
Decreased skin temperature
Mobilization of blood sugar
Increased in blood fatty acids level
Arousal or relaxation

Cellular or Molecular Effects

Increased synthesis/release of hormones
Activation of thyrosine hydroxylase enzyme
Activation of certain transcription factors
Induction of heat shock proteins
Induction of oxidative stress
Reduction of apoptosis
Induction of chromosomal aberrations

Adapted from Yildiz[4]

sical psychomotor responses usually associated with psychostimulants. In concomitant exposure to psychoactive drugs such as ethanol and morphine, nicotine acts as an active enhancement to stimulus discrimination and memory.[12,13] The wide range of effects produced by nicotine is, in part, the result of the ubiquitous distribution of nicotinic receptors in the brain, the heterogeneity of the various nicotinic receptor structures, and its diverse configuration.[14]

THE NICOTINIC RECEPTOR

Neuronal nicotinic acetylcholine receptors (nAChRs) are acetylcholine (ACh)-gated ion channels which belong to a super family of ligand-gated channels that include gamma amino butyric acid (GABA), glycine and 5-hydroxytryptamine (5-HT3) receptors.[15,16] They are composed of varying subunits expressed in mammalian muscle as $\alpha1$, $\beta1$, δ, ε, and γ, and in neurons as $\alpha2 - \alpha7$, $\beta2 - \beta4$, and as $\alpha9$, $\alpha10$ in the epithelium, each having distinctive pharmacological and physiological characteristics.[17,18] The nAChRs are distributed throughout the CNS and may be found in the magnocellular basal complex, peduncolopontine-laterodorsal tegmental complex, striatum, lower brain stem, and the habenula-inter-peducular system.[3,19] The global location, subunit configuration, and interactions with transmitter system make the study of the nAChRs complex at best and confounding in many cases.[20] Generally, the nicotinic receptor sequence consists of: (1) a large hydrophilic amino terminal domain, (2) a compact hydro-phobic domain split into three segments of amino acids termed M1-M3, (3) a small highly variable hydrophilic domain and (4) a hydro-phobic C terminal domain of approximately 20 amino acids termed M4[21] (Figure 1).

Contributing to the depth of nicotine's effects are the neuronal locations of nAChR sites. nAChRs may be found on the pre-, post- and perisynaptic (axonal) areas of neurons. The lo-

FIGURE 1

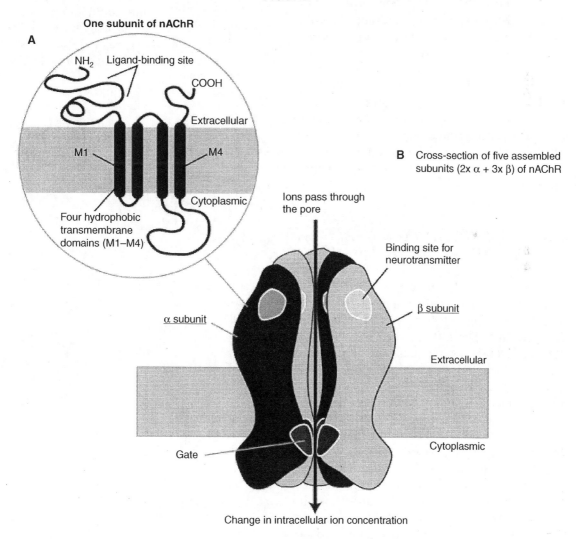

(A) One subunit of the nAChR comprises: (1) an N-terminal extracellular domain, which is involved in binding to neurotransmitters, (2) four hydrophobic transmembrane domains called M1, M2, M3 and M4, and (3) a long cytoplasmic loop between M3 and M4, and other shorter loops connecting the domains.
(B) In cross-section (bottom right), an assembled nAChR has five subunits, each with a binding site and a gate region. All ligand-gated ion channels bind specific neurotransmitters (in this case, nicotine), which induce a conformational change in the receptor allowing the channel opening.

Reprinted from *Progress in Neurobiology*, volume 61, by David Paterson and Agneta Nordberg, titled: Neuronal nicotinic receptors in the human brain, pages 75-111, copyright 2000, with permission from Elsevier.

cation of the receptor site regulates the action of the cell.[18] For example, neuronal presynaptic sites are known to regulate transmitter secretion such as noradrenalin.[22] Postsynaptic receptor sites are considered to regulate fast synaptic transmission and possibly have neuroprotective qualities (please see Gotti and Clementi[19] for a more detailed review). The neuronal nAChR structure, unlike that of the muscle receptor, exists in homo- and heteropentametric configurations of α alone and α and β combination subunits.[16] In the human brain, heteromeric $\alpha 4\beta 2$ and homomeric $\alpha 7$ configurations are thought to exist and are targets of much research interest. Sixteen functional nicotinic receptors configurations have been identified.[19]

Finally, nAChRs are subject to allosteric modulation from a variety of compounds that do not bind to classical ACh sites. These modulating ligands are noncompetitive with nAChR agonists and antagonists but do act to either enhance or reduce ion conductance and channel opening.[14]

AGONISTS, ANTAGONISTS AND OTHER NICOTINIC LIGANDS

Nicotinic AChRs may exist in different states,[14] with at least four distinct functional states at any time. In terms originally suggested by Katz and Thesleff, these conformational states are: (1) a resting state, (2) an activated state, (3) refractory states that are milliseconds or (4) minutes in length.[14,23,19] Binding of nicotine and nicotinic agonists stabilizes the open and desensitized states. The nAChR is permeable to the ions Na^+ and K^+ (postsynaptic) and to Ca^{2+} (presynaptic).[20] Opening and closing of the membrane pores are normally rapid, and as receptor agonists dissipate, the membrane tends to remain open, usually for about 1 millisecond.[17] Additionally, nAChRs are suggested to open sometimes with only one agonist bound and more rarely in the absence of agonist binding, although they may close spontaneously even when ACh is bound.[24] ACh binding only creates a probability of pore opening, which increases as more ACh binds.[25] Finally, nAChRs act diametrically to reactions of other ionic re-

ceptors in the presence of chronic agonist exposure. In a paradigm of chronic exposure to agonists including nicotine, nAChR sites are thought to up-regulate through post-transcription.[19] This atypical response is putatively a compensatory reaction to the rapid desensitization and inactivation during chronic agonist exposure.[26] Nicotinic agonists, antagonists, activators, and inhibitors (Figure 2, Table 2) produce their effects by action at respective ligand-binding sites.

A more complete review of the complexity of the neuronal nicotinic acetylcholine receptor, its agonists, antagonists, activators, inhibitors, multiple structure types, and functions is beyond the scope of this article but may be found in Siegel et al.[16]

FIGURE 2. Structures of various nicotinic cholinergic agonists, antagonist, activators, inhibitors, and blockers.

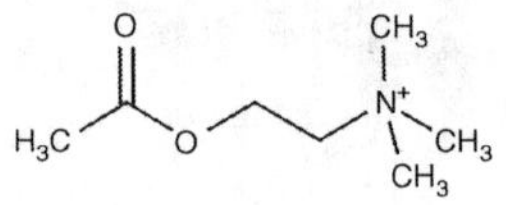

Acetylcholine

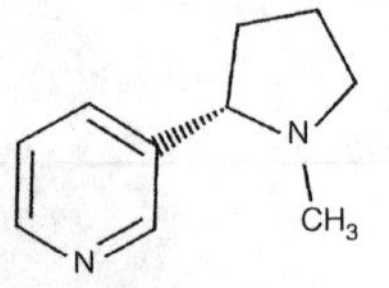

Nicotine, nAChR agonist

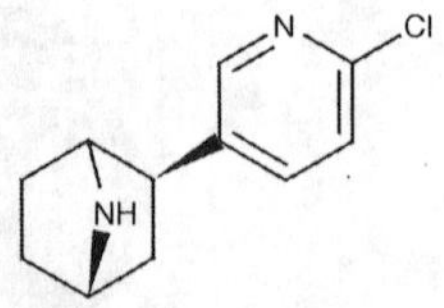

Epibatidime, nAChR agonist

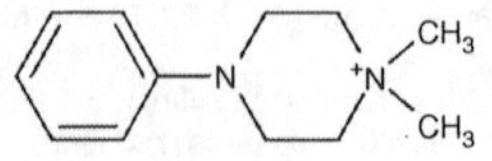

1,1-dimethyl-4-phenylpiperazinium, nAChR agonist (DMPP)

(S)-3-methyl-5-(1-methyl-2-pyrrolidinyl) isoxazole, nAChR agonist (ABT-418)

Trimethaphan, nAChR competitive antagonist

d-Tubocurarine, nAChR antagonist

Galanthamine, nAChr allosteric activator

Phencyclidine, nAChR allosteric inhibitor

Mecamylamine, nAChR channel blocker

Structures courtesy of the U.S. National Library of Medicine.

BEHAVIORAL EFFECTS

Nicotine acts as an effective and powerful agonist of several subpopulations of nicotinic receptors in cholinergic pathways.[14] Acute doses produce alteration of mood, although chronic use produces substantially less sensitivity to such effects suggesting that tolerance develops to some of the effects.[19] A dose-related psychoactive effect in humans elevates scores on standardized tests for liking and euphoria indicating a potential for dependence similar to the effects of stimulants.[26] The potential for dependence associated with nicotine use seems equal to that of other commonly abused psychoactive substances. In animal models, nicotine has been observed to serve as a potent reinforcer. It induces intravenous self-administration, facilitates intracranial self-stimulation, conditioned place preference and facilitates discriminative stimulus tasks.[26] Research concerning nicotine's tendency to relieve negative emotion has produced a number of additional questions. For example, Kassel, Stroud, and Paronis found that there is very little known about stress reduction among those initiating smoking, regular smokers, and the negative emotion of smokers attempting to quit.[27] Using cardiac function as a measure of stress, abstinent smokers became more stressful when completing mental arithmetic tasks, yet during stressful circumstances, smoking exacerbated stress.[28] Conversely, when stress was measured by corticosteroid levels, psychostimulants (amphetamine and nicotine) were found to reduce measurable stress in female rats.[29]

MEMORY

A number of studies have implicated nicotine as an enhancer of memory as well as learning and attention.[30-32] Cholinergic projections from the septal area innervate the hippocampus.[33] In the hippocampus, the antagonist scopolamine, a muscarinic antagonist, significantly impairs memory performance. This leads investigators to believe cholinergic muscarinic receptors play the major role in memory.[34] The nAChRs are thought to be a factor in modulating hippocampal activity, in part by the activation of GABAergic interneurons.[29] The com-

TABLE 2. Partial list of neuronal nicotinic agonists, antagonists, activators, inhibitors, and blockers

Agonist	Antagonist	Allosteric Activators	Allosteric Inhibitors	Channel Blockers
ACh	DHβE	Physostigmine	Ethanol	Mecamylamine
Nicotine	Methylcaconitine	Galanthamine	Phencyclidine	Chlorisondamine
Epibatidine	d-Tubocurarine	Tacrine	MK801b	Hexamethonium
ABT-418	α-Bungarotoxin	Benzoquinonium	Chlorpromazine	
Cytisine	n-Bungarotoxin	Codeine	Progesterone	
GTS-21	Strychnine	5-HT	Corticosterone	
Anabaseine			Dexamethasone	

ABT-418, 3-Methyl-5-[(2S)-1-methyl-2-pyrrolidinyl]isoxazole hydrochloride; **DHβE**, Dihydro-β-erythroidine; **GTS-21**, [3-(2,4-Dimethoxybenzylidene)-anabaseine.

mon nAChR types found in the hippocampus are the $\alpha4$, $\alpha2$, and $\alpha7$.[35]

Puma et al. have demonstrated that nicotine produces an improvement of retention in animals performing a novel-familiar object discrimination task.[36] This memory enhancement was evidenced whether nicotine was administered pretrial, posttrial, or pretest suggesting nicotine may enhance the acquisition as well as the restitution and retention of information.

Xu and colleagues describe the memory-enhancing aspects of nicotine as reducing latency of recall.[37] Increased item recall in nicotine-satiated smokers relative to abstinent smokers has also been observed. The researcher's found that in satiated smokers, activity increased in the left dorsal lateral prefrontal cortex (left DLPFC) suggesting that working-memory is related to the excitation of dopaminergic pathways of the striatum and recall processes associated with the left DLPFC. Some investigators suggest that the abnormal activity associated with working memory in the DLPFC of schizophrenics indicates impairment of processes involved in working memory in this region.[38] Working memory and other cognitive functions are often impaired in schizophrenia.[39] The cognitive-enhancing effects of nicotine and nAChr inhibitors have had limited success in the treatment of perseveration often associated with attention deficit and schizophrenia.[40,41] However, Kumari and Postma have suggested high prevalence rates of smoking in the population suffering schizophrenia represents a form of self-medicating.[42] The benefits of smoking (nicotine) may be explained in terms of nicotine's interaction with dopaminergic and glutamatergic transmitter systems and pathways. Smoking may be used by schizophrenics as a form of self-medication, reducing cognitive and sensory deficits, reduced side-effects of antipsychotic medications and psychiatric symptoms.[42]

Direct stimulation of the nAChRs with nicotine in combination with vitamin E has been shown to enhance memory in animals with ACh-impaired functioning as may be seen in Alzheimer's disease.[43] These researchers have demonstrated that the blockade of nicotinic receptors by mecamylamine reduced memory retention. Activation of these receptors by nicotine is suggested to increase retention of nAChRs. A loss of nicotinic binding sites is associated with plaques and tangles found in Alzheimer's disease. Increased activities of acetylcholine, dopamine, norepinephrine, serotonin and glutamate have been associated with nicotine and may be related to improved memory.[43]

The complexity of the relationship of nicotinic receptors and nicotine-enhanced memory has been discovered through research by Hefco et al.[44] These researchers found nAChRs activated by nicotine improved short-term memory in normal rats and ameliorated impaired memory induced by nAChR and dopamine D2 blockade. Improvements in long-term memory were not found in acute administration of nicotine; however improvement in performance was established in chronic nicotine administration. The variation of effect of acute versus chronic doses of nicotine on short- and long-term memory may be attributed to the fact that different brain regions are involved in storage and retrieval in these two categories of memory. The performance increase in animals with blocked nAChRs seems to indicate that not all acetylcholine receptors are blocked by certain drugs and that these available receptors play a role in memory as well.[44]

Activation of presynaptic nicotinic receptors is suggested to modulate glutamatergic as well as GABAergic synaptic transmission in the amygdale.[45] Two separate and distinct cholinergic mechanisms in the basolateral complex of the amygdala were found to interact: a muscarinic system that modulates only working memory (5s in duration) and long-term memory (24 h in duration), and a nicotinic system that involves working memory, short-term memory (90 min in duration) and long-term memory. This contribution of nAChRs in the amygdala is particularly interesting because nicotine, acting as a positively reinforcing stimulus, can affect several cognitive functions, such as attention and learning.[46] Furthermore, findings that relate reward to attention and memory may provide clues to the processes involved in addiction.[45]

ATTENTION

Closely associated with memory, attention is considered to involve the capacity to disengage a stimulus to shift focus and response to either sensory or semantic stimulus characteristics.[47] Animal experiments designed to discover the effect of nicotine on attention found that nicotine increased the accuracy of subject responses and the magnitude of the accuracy-effect was dose-related to nicotine.[8] These researchers concluded that nicotine can produce robust performance in normal rats in tasks requiring attention. More precisely, the investigators suggested that nicotine enhanced vigilance. Furthermore, the enhancement in performance excluded changes in sensory or motor capability, learning or memory. Grilly, Simon, and Levin found that nicotine significantly improved stimulus detection accuracy while decreasing choice latency in a dose-dependent manner among young and old rats.[6] These results were similar to human responses to nic-

otine treatment. This suggests that improved performance in tasks heavily dependent on attention are more likely to be a result of heightened attention to visual stimuli rather than more complex processes such as memory, learning, or discrimination.

More recent research regarding nicotine's impact on attention is less equivocal and has highlighted other contributions of nicotine. Specifically, Lawrence, Ross, and Stein noted, "Nicotine induced enhancement of activation in smokers, regardless of treatment order, reinforces the notion that nicotine facilitated sustained attention" (p. 543).[48] Furthermore, these researchers noted that all known cholinergic sites (thalamus, caudate, substantia nigra, and to a lesser extent the parietal and occipital, frontal, and temporal cortices) were activated by nicotine. Activation of the parietal cortex is associated with enhanced attention, and the thalamus and caudate with general increases in arousal and motor activity (see Lezak et al.[47] for a more complete discussion). These findings are consistent with current thoughts on attention-deficit disorders in that cholinergic enhancement seems to increase the signal-to-noise ratio (decrease distraction) of the environment.[49]

Contradictory to the above findings is the oft-cited review by Heishman, Taylor, and Henningfield.[32] With the exception of attention, they noted that cognitive functioning was not enhanced by nicotine uniformly across all participants. They also noted that the 101 studies reviewed contained methodological problems such as imprecise dosing, lack of single- or double-blind experimental conditions, and ad lib cigarette use in many of the reviewed articles.

ATTENTION DEFICIT DISORDER, ALZHEIMER'S, AND PARKINSON'S

Considering the improvement in attention given to nAChR stimulation, there exist implications for nicotine or nicotinic agonists as a treatment for Attention Deficit/Hyperactivity Disorder (ADHD). The Diagnostic and Statistical Manual of Mental Disorders considers nine criteria for consideration in diagnosing ADHD.[50] Included in the criteria are failure to attend details and lack of sustained attention. ADHD has been associated with brain activity that is modulated by catecholamines, which are affected by stimulant drugs.[51] Specifically, methylphenidate, amphetamine, and amphetamine mixtures are currently used in children and adults.[52] In addition to these stimulants, nicotine has been studied as a therapeutic agent for the treatment of ADHD.[53] The symptomatology of ADHD often leads to difficulty in school and other behavioral problems which has raised debate whether smoking is an act of self-medication. Pomerleau et al. found that 40% of adults with ADHD smoked cigarettes compared to 26% in the general population.[54]

Bekker et al. reported that acute nicotine effects were apparent for attention-related measures and reflected improved attention.[49] The effects were more apparent in those suffering ADHD rather than in chronic nicotine abuse. In a study of smoking and nonsmoking sufferers of ADHD, Levin et al. found overall nicotine-induced improvement in mood and affect in nonsmokers, signifying these improvements are not simply due to relief from nicotine withdrawal of smokers.[53]

The key features of Alzheimer's disease are a loss of cognitive function including short-term memory, impaired attention associated with restlessness, language disturbances, and emotional instability. Although many transmitter systems are thought to be involved, nicotinic mechanisms have been reported to explain the pathology and to aid in the design of treatment for Alzheimer's disease.[55] Modulation of nicotinic receptors is thought to be a means of producing a therapeutic benefit for sufferers of Alzheimer's.[55,56]

An early report of the use of nicotine patches on Alzheimer's patients found that learning trial scores increased in those patients using the nicotine patch versus placebo.[57] Interestingly, the learning effects persisted beyond the active drug trials. This suggests an increase in nAChR activity. Whether these and similar results are through increased activity of existing receptors or the generation of new receptors through gene expression is yet unanswered.

Although not usually associated with nAChRs, Parkinson's disease has become a target of nAChR research.[58-60] Degeneration associated with dopaminergic activity in the nigrostriatal system has also been related to declining nAChR activity.[61] However, stimulation of the $\alpha 7$ subunits in this area by nicotine has partially protected surrounding dopaminergic neurons.[62] In a review, Fratiglioni and Wang found that while both incidence and mortality rates of idiopathic PD were detected among smokers and nonsmokers, the incidence for PD in smokers was less than half that in nonsmokers and the lower incidence was present in all age strata.[63] Alterations in the nAChr subtypes and well as their expression in the nigrostriatal system have been suggested as a cause of reduced control of motor activity in Parkinson's.[64] The cumulative evidence suggests that stimulation of nAChRs in the basal ganglia results in functional improvements at the cellular-level. These changes may include increased control of motor activity and a neuroprotective quality (see Paterson and Nordberg[14] for a detailed review).

NICOTINE REPLACEMENT THERAPY: DEMOGRAPHICS AND RESPONSE RATES AND IMPACT OF ADDICTION AND PSYCHIATRIC DISORDERS

Despite the widespread admonitions of healthcare professionals, the Federal Government, and the eventual admissions of tobacco manufacturers regarding the dangers of smoking, tobacco use is estimated to be the single largest cause of premature death.[65] The addictive nature of tobacco is primarily due to nicotine which is known to act by stimulating the ventral tegmental area (VTA) and the nucleus accumbens. Physiologically relevant amounts of nicotine have been shown to activate pre-, post-, and perisynaptic nAChRs in the VTA.[14,18,23]

Survey studies of adolescent smoking suggest nicotine dependence may occur after only a few cigarettes, which are consistent with research reports of the addictive power of nicotine.[65,66] Once addicted, most individuals find stopping exceptionally difficult. A physiological explanation may lie in the number of different transmitters affected by nicotine.[14] Other aspects of smoking have been considered to better grasp the totality of smoking addiction. In addition to the neurobiological theories of addiction are psychosocial smoking (social psychology), sensorimotor smoking (classical and operant conditioning), stimulation and sedation smoking (arousal and affect regulation), dependent smoking (avoidance of withdrawal), and finally, automatic smoking (learning and neurobiological).[67]

The most common form of pharmacological treatment for addiction to tobacco products is nicotine replacement therapy (NRT).[68] Introduced in 1984, NRT was intended to replace blood nicotine and hence assuage physiological withdrawal symptoms (e.g., irritability and depressed mood) and cigarette cravings experienced by most smokers on attempts at quitting cigarettes.[69] The course of therapy typically ranges from 1-3 months for most NRT products.[70] NRT is considered safer than the use of tobacco products because it does not contain carcinogens or other toxic chemicals found in tobacco smoke, and there is a reduced risk of addiction to and withdrawal from NRT.[71] When combined with psychoeducational counseling, NRT has been proven effective for tobacco use cessation. Many "starter kits" of NRT products contain self-help information intended to provide means of educational insight.

Anderson reviewed the effectiveness of NRT/bupropion compared to a non-drug intervention.[71] Significant differences existed at baseline for age groups and NRT/bupropion and were used as covariates in a logistic regression analysis to predict forward movement toward smoking cessation. The research sample's demographics and statistical results are presented in Table 3.

A similar study by John et al. examined a sample of German smokers.[72] They concluded that gender and age group were not important in proceeding through the transtheoretical model's stages of change. Of all factors examined, use of NRT yielded the largest effect size (0.16). A summary of the relevant results is contained in Table 4.

In a sample of veteran inpatients being treated for alcohol and drug dependence, the opportunity to begin NRT (nicotine patch) was openly offered.[73] Of the patients attempting cessation, 6.1% (3) were abstinent at day 21 and 10.2% (5) continued to seek prescription transdermal patches as outpatients. All five subjects had a diagnosis of alcohol dependence; one also had a diagnosis of cocaine dependence. One patient had a psychiatric diagnosis of post-traumatic stress disorder and depression (Table 5).

For the years 1992 through 1997, the CDC estimated the use of nicotine in NRT formulations increased to 7.7 million attempts annually, and Hughes et al. found that OTC formulations were equally efficacious as prescription formulations.[74,75] This being stated, there is also evidence of off-label use of NRT preparations through persistent use beyond six months and concomitant use of NRT nasal spray and smoking.[76,77] In both cases, studies indicate misuse is transient and involves a relatively low percentage of users for gum and patches and a somewhat higher rate for inhalers and nasal spray.

CONCLUSIONS

The pharmacology of naturally-occurring and synthetic nicotinic receptor agonists/antagonists/activators/inhibitors suggests that selective nAChR subtype activation may have considerable potential in the treatment of human neurological disorders such as Parkinson's disease, Alzheimer's disease Tourette's syndrome, attention-deficit hyperactivity, anxiety, pain, addiction liability, memory loss, and feeding to name a few.[3,5-13] However, use of nicotine has been limited thus far because of its adverse effects (heart rate, temperature, respiration, and addiction potential) and due to the challenges presented by the complexity and diversity of nAChRs.[4]

Although the effects of nicotine on the central nervous system are not fully understood, they are contributing to a better understanding of neurotransmitter systems and some pathological states. Therefore, more intensive study of the pharmacodynamics of nicotine and nicotinic agonists may help improve quality of life of those suffering degenerative diseases. The use of structure-activity studies taking natural nicotinic agonists as the starting point are revealing some rules for enhancing agonist potency and application, although these rules are often not intuitive. There is evidence that a synergistic effect of nicotine on the release of GABA and monoamines such as norepinephrine, dopamine, serotonin, and acetylcholine

TABLE 3. Sample Characteristics

Variable		%	N	*N*
Age group (years)	18-25	0.8	1	
	26-35	5.8	7	
	36-45	14.2	17	
	46-55	34.2	41	
	56-65	34.2	41	
	66-75	10.8	13	
	Total			120
Income	< 15K	27.5	33	
	15-25K	24.2	29	
	25,001-35K	16.7	20	
	35,001-45K	9.2	11	
	> 45K	20.0	24	
	Total			117
Ethnicity	African American	8.3	10	
	Caucasian	81.7	98	
	Hispanic	6.7	8	
	Other	3.3	4	
	Total			120
Gender	Male	65.8	79	
	Female	34.2	41	
	Total			120
Education	< 9th Grade	3.3	4	
	Grade 9-11	14.2	17	
	High School/GED	39.2	47	
	College or higher	42.5	51	
	Total			119
Age of smoking onset	6-10	3.3	4	
	11-15	30.0	36	
	16-20	45.8	55	
	21-25	11.7	14	
	26-30	5.0	6	
	31-35	0.8	1	
	36 and up	1.7	2	
	Total			118

Summary, logistic regression analysis predicting forward movement in stages of change			
Variable	*B*	SE *B*	*p*
Age	0.02	0.03	0.43
Onset of smoking	0.07	0.50	0.17
Gender	0.09	0.59	0.88
NRT/bupropion	1.20	0.38	0**

**Statistically significant, *p* < .001

Adapted from Anderson[71]

TABLE 4. Sample Characteristics

Variable		%	N	N
Age group (years)	18-24	11.1	119	
	25-44	56.0	602	
	45-64	32.7	352	
	Total			1073
Income (Monthly)	≤ 1,030*	13.5	145	
	> 1,030*	81.2	873	
	Total			1018
Ethnicity	African American	8.3	10	
	Caucasian	81.7	98	
	Hispanic	6.7	8	
	Other	3.3	4	
	Total			120
Gender	Male	50.3	541	
	Female	49.7	534	
	Total			1075
Education	< 12th Years	9.7	104	
	12 Years	43.8	471	
	13-15 Years	40.6	436	
	≥ 16 Years	6.0	64	
	Total			1075

Use of NRT	Stage of Change			
	Precontemplation $N = 821$ %n	Contemplation $N = 183$ %n	Preparation $N = 71$ %n	Effect size**
Yes	57.1	30.3	12.6	
No	78.8	15.4	5.9	0.16

* Euros converted to USD for mid-2003.
** Cohen's ω^2

Adapted from John, Meyer, Rumpf, and Hapke[72]

may have a strong positive effect on mood and enhanced cognitive functioning.[48,51,78] The most recent research on nicotine has focused on its potential use as a therapeutic agent.[18] The results of clinical trials currently being conducted in different neurological and psychiatric diseases will be of great importance for future therapeutic use of nicotine as well as for the understanding of nicotine actions within the brain.[7,10,18] With the understanding gained from the study of nicotine, nicotinic agonists are being discovered which may be used as therapeutic agents in the pathological conditions presented in this review.[18,19]

TABLE 5. Sample Characteristics

Variable		%	N	N
Primary substance dependence diagnosis	Alcohol dependence	79.6	39	
	Cocaine dependence	12.2	6	
	Polysubstance dependence	8.2	4	
	Total			49
Secondary substance dependence diagnosis	None	51.0	25	
	Alcohol dependence	8.2	4	
	Cocaine dependence	28.6	14	
	Opioid dependence	6.1	3	
	Cannabis dependence	4.1	2	
	Amphetamine dependence	2.0	1	
	Total			49
Primary psychiatric diagnosis	None	38.8	19	
	Depression	26.5	13	
	Bipolar	2.0	1	
	PTSD	16.3	8	
	Panic disorder	2.0	1	
	Other	14.3	7	
	Total			49
Secondary psychiatric diagnosis	None	81.6	40	
	Depression	10.2	5	
	PTSD	8.2	4	
	Total			49
Ethnicity	African American	22.4	11	
	Caucasian	75.5	37	
	Native American	2.0	1	
	Total			49
Gender	Male	95.9	47	
	Female	4.1	2	
	Total			49

Adapted from Saxon, McGuffin, Walker[73]

REFERENCES

1. Rafferty SM. Evidence of early tobacco in Northeastern North America. Journal of Archeological Science (in press) 2006.

2. Tariq M, Khan HA, Elfaki I, Al Deeb AI Moutaery K. Neuroprotective effect of nicotine against 3-nitropropionic acid (3-NP)-induced experimental Huntington's disease in rats. Br Res Bull. 2005; 67: 161-168.

3. Domino EF. Nicotine and tobacco dependence: Normalization or stimulation. Alcohol 2000; 24(2): 83-86.

4. Yildiz D. Nicotine, its metabolism and an overview of its biological effects. Toxicon. 2004; 43:619-632.

5. Luine VN, Mohan G, Tu Z, Efange SMN. Chromaproline and chromaperidine, nicotine agonists,

and Donepezil, cholinesterase inhibitor, enhance performance of memory tasks in ovariectomized rats. Pharm Biochem Behav 2002; 74:213-220.

6. Grilly DM, Simon BB, Levin ED. Nicotine enhances stimulus detection performance of middle- and old-aged rats: A longitudinal study. Pharmacol Biochem Behav. 2000; 65:665-670.

7. Sherwood N. Effects of nicotine on human psychomotor performance. Hum Psychopharmacol. 1993; 8:155-184.

8. Stolerman IP, Mirza NR, Hahn B, Shoaib M. Nicotine in an animal model of attention. Eur J Pharmacol. 2000; 393:147-154.

9. Stevens SL, Colwell B, Smith DW, Robinson J, McMillan C. An exploration of self-reported negative affect by adolescents as a reason for smoking: Implications for tobacco prevention and intervention programs. Prev Med. 2005; 41:589-596.

10. Grottick AJ, Wyler R, Higgins GA. A study of the nicotinic agonist SIB-1553A on locomotion, and attention as measured by the five-choice serial reaction time task. Pharmacol Biochem Behav. 2001; 70:505-513.

11. Grilly DM. A verification of psychostimulant-induced improvement in sustained attention in rats: Effects of d-amphetamine, nicotine, and pemoline. Exp Clin Psychopharmacol. 2000; 8:14-21.

12. Le Foll B, Goldberg SR. Ethanol does not affect discriminative-stimulus effects of nicotine in rats. Eur J Pharmacol. 2005; 519: 96-102.

13. Zarrindast M, Fattahi Z, Rostami P, Rezayof A. Role of the cholinergic system in the rat basolateral amygdala on morphine-induced conditioned place preference. Pharmacol Biochem Behav. 2005; 8:1-10.

14. Paterson D, Nordberg A. Neuronal nicotinic receptors in the human brain. Prog in Neurobiol 2000; 61:75-111.

15. Karlin A, Akabas MH. Toward a structural basis for the function of nicotinic acetylcholine receptors and their cousins. Neuron 1995; 15:1231-1244.

16. Siegel GJ, Agranoff BW, Fisher SK, Albers R, Uhler MD. Basic Neurochemistry: Molecular, Cellular and Medical Aspects, Sixth Edition. Philadelphia: Lippincott Williams and Wilkins, 1999.

17. Galzi JL, Changeux JP. Neuronal nicotinic receptors: Molecular organization and regulations. Neuropharmacology 1995; 34: 63-582.

18. Cassels BK, Bermúdez I, Dajas F, Abin-Carriquiry JA, Wonnacott S. From ligand design to therapeutic efficacy: The challenge for nicotinic receptor research. Drug Discov Today 2005; 1023-24:1657-1665.

19. Gotti C, Clementi F. Neuronal nicotinic receptors: From structure to pathology. Progressive Neurobiology 2004; 74:363-396.

20. Itier V, Bertrand D. Neuronal nicotinic receptors: From protein structure to function. Federation of European Biochemical Societies 2001; 504(3): 118-125.

21. Brown DA. The history of the neuronal nicotinic receptor. In: Clementi F, Fornasari D, Gotti C, ed.

Neuronal nicotinic receptors. New York: Springer 2000:3-11.

22. Wonnacott S. Presynaptic nicotinic ACh receptors. Trends Neurosci. 1997;20(2):92-98.

23. MacDermott AB, Role LW, Siegelbaum SA. Presynaptic ionotropic receptors and the control of transmitter release. Ann Rev Neurosci. 1999; 22:443-485.

24. Katz B, Thesleff S. A study of the desensitization produced by acetylcholine at the motor end plate. J Physiol. 1957; 138:63-80.

25. Colquhoun D, Sivilotti LG. Function and structure in glycine receptors and some of their relatives. Trends Neurosci. 2004; 27:337-344.

26. Di Chiara G. Behavioral pharmacology and neurobiology of nicotine reward and dependence. In: Clementi F, Fornasari D, Gotti, C, ed. Neuronal nicotinic receptors. New York: Springer 2000:603-750.

27. Kassel JD, Stroud LR, Paronis CA. Smoking, stress, and negative affect: Correlation, causation, and context across stages of smoking. Psychol Bull. 2003; 129: 270-304.

28. VanderKaay MM, Patterson SM. Nicotine and acute stress: Effects of nicotine versus nicotine withdrawal on stress-induced hemoconcentration and cardiovascular reactivity. Biol Psychol 2006; 71(2):191-201.

29. McCormick CM, Robards D, Kopeikina K, Kelsey JE. Long-lasting, sex- and age-specific effects of social stressors on corticosterone responses to restraint and on locomotor responses to psychostimulants in rats. Horm Behav. 2005; 48: 64-74.

30. Jones S, Sudweeks S, Yakel JL. Nicotinic receptors in the brain: Correlating physiology with function. Trends Neurosci. 1999; 22: 555-561.

31. Brioni JD, Decker MW, Sullivan JP, Arneric SP. The pharmacology of (-) nicotine and novel cholinergic channel modulators. Adv Pharmacol. 1997; 37:153-214.

32. Heishman SJ, Taylor RC, Henningfield JE. Nicotine and smoking: A review of effects on human performance. Exp Clin Psychopharmacol 1994; 2:345-395.

33. Levin ED. The role of nicotinic acetylcholine receptors in cognitive function. In: Clementi F, Fornasari D, Gotti, C, ed. Neuronal nicotinic receptors. New York: Springer 2000:587-602.

34. Kim JS, Levin ED. Nicotinic, muscarinic and dopaminergic actions in the ventral hippocampus and the nucleus accumbens: Effects on spatial working memory in rats. Brain Res. 1996; 725:231-240.

35. Felix R, Levin ED. Nicotinic antagonist administration into the ventral hippocampus and spatial working memory in rats. Neuroscience 1997; 81:1009-1017.

36. Puma C, Deschaux O, Molimard R, Bizot JC. Nicotine improves memory in an object recognition task in rats. Eur Neuropsychopharmacol. 1999; 9:323-327.

37. Xu J, Mendrek A, Cohen MS, Monterosso J, Rodriguez P, Simon SL, Brody A, Jarvik M, Domier CP, Olmstead R, Ernst M, London E. Brain activity in cigarette smokers performing a working memory task: Effect of smoking abstinence. Biol Psychiatry. 2005;58: 143-150.

38. Perlstein WM, Dixit NK, Carter CS, Noll DC, Cohen JD. Prefrontal cortex dysfunction mediates deficits in working memory and prepotent responding in schizophrenia. Biol Psychiatry. 2003; 53:25-38.

39. Stubotnik KL, Nuecterlein KH, Green MF, Horan WP, Nienow TM, Ventura J, Nguyen A. Neurocognitive and social cognitive correlates of formal thought disorder in schizophrenia patients. Schizophr Res 2006;85: 84-95.

40. Poltavski DV, Petros T. Effects of transdermal nicotine on attention in adult non-smokers with and without attentional deficits. Physiol Behav. 2006; 87: 614-627.

41. Schubert MH, Young KA, Hicks PB. Galantamine improves cognition in schizophrenia patients stabilized on risperidone. Biol Psychiatry. 2006; 60(6):530-533.

42. Kumari V, Postma P. Nicotine use in schizophrenia: The self medication hypothesis. Neuroscie Behav Revs. 2006; 29:1021-1034.

43. Eidi A, Eidi M, Mohmoodi G, Oryan S. Effect of vitamin E on memory retention in rats: Possible involvement of cholinergic system. Eur Neuropsychopharmacol. (in press), 2006.

44. Hefco V, Yamada K, Hefco A, Hritcu L, Tiron A, Olariu A, Nabeshima T. Effects of nicotine on memory impairment induced by blockade of muscarinic, nicotinic and dopamine D2 receptors in rats. Eur J Pharmacol. 2003; 474:227-232.

45. Barros DM, Ramirez MR, Izquierdo I. Modulation of working, short-term and long-term memory by nicotinic receptors in the basolateral amygdala in rats. Neurobiol Learn Mem. 2005; 83:113-118.

46. Maren S. Long-term potentiation in the amygdala: A mechanism for emotional learning and memory. Trends Neurosci. 1999; 561-567.

47. Lezak MD, Howieson DB, Loring DW. Neuropsychological Assessment. 4th ed. New York: Oxford Press, 2004.

48. Lawrence NS, Ross TJ, Stein EA. Cognitive mechanisms of nicotine on visual attention. Neuron 2002; 36:539-548.

49. Bekker EM, Böcker KBE, Van Hunsel F, van den Berg MC, Kenemans JL. Acute effects of nicotine on attention and response inhibition. Pharmacol Biochem Behav. 2005; 85:539-548.

50. American Psychiatric Association. Diagnostic and Statistical Manual of Mental Disorders (4th Ed.). Washington, DC, 2000.

51. Swanson J, Castellanos FX, Murias M, LaHoste G, Kennedy J. Cognitive neuroscience of attention deficit hyperactivity disorder and hyperkinetic disorder. Curr Opin Neurobiol. 1998; 8:263-271.

52. Reitz SJ. Attention-deficit/hyperactivity disorder: Focus on pharmacologic management. Journal of Pediatric Health Care. 1997; 11:78-83.

53. Levin ED, Conners CK, Sparrow E, Hinton SC, Erhardt D, Meck WH, Rose JE, March J. Nicotine Effects on adults with attention-deficit/hyperactivity disorder. Psychopharmacol. 1996; 123(1):55-63.

54. Pomerleau OF, Downey KK, Stelson FW, Pomerleau CS. Cigarette smoking in adult patients diagnosed with attention deficit hyperactivity disorder. J Subst Abuse. 1995; 7(3):373-378.

55. Salamone F, Zhou M, Auerbach A. A re-examination of adult mouse nicotinic acetylcholine receptor channel activation kinetics. Am J Physiol. 1999;15:516 (Pt 2):315-330.

56. Newhouse PA, Potter A, Kelton M, Corwin J. Nicotinic treatment of Alzheimer's disease. Biol Psychiatry. 2000; 49(3):268-278.

57. Wilson AL, Langley LK, Monley J, Bauer T, Rottunda S, McFalls E, Kovera C, McCarten JR. Nicotine patches in Alzheimer's disease: A pilot study on learning, memory and safety. Pharmacol Biochem and Behav. 1995; 51(2-3):509-514.

58. Whitehouse PJ, Hedreen JC, White CL, Price DL. Basal forebrain neurons in the dementia of Parkinson disease. Ann Neurol. 1983; 13:243-248.

59. Aubert I, Araujo DM, Cecyre D, Robitaille Y, Gauthier S, Quirion R. Comparative alterations of nicotinic and muscarinic binding sites in Alzheimer's and Parkinson's diseases. J Neurochem. 1992; 58:529-541.

60. Lange KW, Wells FR, Jenner P, Marsden CD. Altered muscarinic and nicotinic receptor densities in cortical and subcortical brain regions in Parkinson's disease. J Neurochem. 1993; 60:197-203.

61. Dluzen DE, Anderson LI. The effects of nicotine on dopamine and DOPAC output from rat striatal tissue. Eur J Pharmacol. 1998; 341:23-32.

62. Jeyarasasingam G, Tompkins L, Quik M. Stimulation of non-a7 nicotinic receptors partially protects dopaminergic neurons from 1-methyl-4-phenylpyridinium-induced toxicity in culture. Neuroscience 2002; 109:275-285.

63. Fratiglioni L, Wang H. Smoking and Parkinson's and Alzheimer's disease: A review of epidemiological studies. Behaviour Brain Res. 2000; 113:117-120.

64. Peto R, Lopez AD, Boreham J, Thun M, Heath Jr C. Mortality from tobacco in developed countries: Indirect estimation from national vital statistics. Lancet 1992; 339: 268- 278.

65. DiFranza JR, Rigotti NA, McNeill AD, Ockene JK, Savageau JA, St Cyr D, Coleman M. Initial symptoms of nicotine dependence in adolescents. Tob Control. 2000; 9:313-319.

66. Kenny PJ, Markou A. Neurobiology of the nicotine withdrawal syndrome. Pharmacol Biochem Behav. 2001; 70:531-549.

67. Lujic C, Reuter M, Netter P. Psychobiological theories of smoking and smoking motivation. European Psychologist. 2005; 10:1-24.

68. Burton SL, Gitchell JG, Shiffman S. Use of FDA-approved pharmacological treatments for tobacco dependence: United States 1984-1998. MMWR 2000; 49:665-668.

69. Hughes JR, Hatsukami DK. Signs and symptoms of tobacco withdrawal. Arch Gen Psychiatr. 1986; 43: 289-294.

70. West R, Hajek P, Foulds J, Nilsson F, May S, Meadows A. A comparison of the abuse liability and dependence potential of nicotine patch, gum, spray, and inhaler. Psychopharmacol. 2000; 149:198-202.

71. Anderson S. Do indigenous helpers foster smoking cessation in adult smokers? Addict Behav. 2006; 31:1496-1502.

72. John U, Meyer C, Rumpf HJ, Hapke U. Relation among stage of change, demographic characteristics, smoking history, and nicotine dependence in an adult German population. Prev Med. 2003; 37:368-374.

73. Saxon AJ, McGuffin R, Walker RD. An open trial of transdermal nicotine replacement therapy for smoking cessation among alcohol- and drug-dependent inpatients. J Subst Abuse Treat. 1997; 14:333-337.

74. Centers for Disease Control and Prevention. MMWR. 2000; 49:29.

75. Hughes JR, Shiffman S, Callas P, Zhang J. A meta-analysis of the efficacy of over-the-counter nicotine replacement. Tob Control. 2005; 14:49-54.

76. Shiffman S, Hughes JR, Pillitteri JL, Burton SL. (2003). Persistent use of nicotine replacement therapy: An analysis of actual purchase patterns in a population based sample. Tob Control. 2003; 12:310-316.

77. Hughes JR, Adams EH, Franzon MA, Maguire MK, Guary J. A prospective study of the off-label use of, abuse of, and dependence on nicotine inhaler. Tob Control. 2005; 14:49-54.

78. Morissette SB, Brown TA, Kamholts BW, Gulliver SB. Differences between smokers and nonsmokers with anxiety disorders. Anxiety Disorders (in press), 2006.

doi:10.1300/J069v25S01_03

Anabolic Steroid Abuse:
Neurobiological Substrates and Psychiatric Comorbidity

Daria Rylkova, BS
Adrie W. Bruijnzeel, PhD
Mark S. Gold, MD

SUMMARY. Anabolic androgenic steroids (AAS) have legitimate medical uses, however during the last decade there has been a strong increase in the abuse of steroids. Numerous physiological and psychological disorders have been associated with the prolonged abuse of AAS. Chronic steroid abuse has been associated with psychiatric disorders such as increased anxiety, depression, mania, and psychosis. In addition, there is evidence of anabolic androgenic steroid addiction among chronic abusers. In recent years, progress has been made in ascertaining the effects of AAS on the central nervous system. AAS have been shown to modulate a wide range of neurotransmitter systems that have been suggested to play a role in brain reward function and in a variety of psychiatric disorders. Further research is warranted to investigate anabolic androgenic steroid addiction liability and to develop novel treatments for withdrawal symptomatology. doi:10.1300/J069v25S01_04 *[Article copies available for a fee from The Haworth Document Delivery Service: 1-800-HAWORTH. E-mail address: <docdelivery@haworthpress.com> Website: <http://www.HaworthPress.com> © 2007 by The Haworth Press, Inc. All rights reserved.]*

KEYWORDS. Anabolic androgenic steroids, abuse, addiction, physiological disorder

INTRODUCTION

Anabolic androgenic steroids (AAS) have legitimate uses in the treatment of hypogonadism, anemia, and growth deficiency as well as for wasting associated with terminal mammary cancer and AIDS.[1-5] However, in recent years the number of people abusing AAS has been increasing. It has been estimated that up to 95% of football players and 80 to 99% of professional body builders abuse AAS.[6] Current assessments of children (12-18 years of age) estimate a 2-4% prevalence of anabolic androgenic steroid abuse.[7] Among non-athletes, anabolic androgenic steroid abuse is particularly common among adolescent boys. It has been

Daria Rylkova is an undergraduate student and Adrie W. Bruijnzeel is Assistant Professor, Department of Psychiatry, University of Florida.

Mark S. Gold is Distinguished Professor, Departments of Psychiatry, Neuroscience, Community Health and Family Medicine, Chief, Division of Addiction Medicine, University of Florida.

Address correspondence to: Adrie W. Bruijnzeel, PhD, University of Florida, Department of Psychiatry, McKnight Brain Institute, 100 South Newell Drive, P.O. Box 100256, Gainesville, FL 32610 (E-mail: awbruijn@psychiatry.ufl.edu).

[Haworth co-indexing entry note]: "Anabolic Steroid Abuse: Neurobiological Substrates and Psychiatric Comorbidity." Rylkova, Daria, Adrie W. Bruijnzeel, and Mark S. Gold. Co-published simultaneously in *Journal of Addictive Diseases* (The Haworth Medical Press, an imprint of The Haworth Press, Inc.) Vol. 25, Supplement No. 1, 2007, pp. 33-45; and: *Performance-Enhancing Medications and Drugs of Abuse* (ed: Mark S. Gold) The Haworth Medical Press, an imprint of The Haworth Press, Inc., 2007, pp. 33-45. Single or multiple copies of this article are available for a fee from The Haworth Document Delivery Service [1-800-HAWORTH, 9:00 a.m. - 5:00 p.m. (EST). E-mail address: docdelivery@haworthpress.com].

estimated that approximately 7% of high school seniors (17-18 years of age) abuse AAS.[8] It is of particular concern that up to two thirds of anabolic androgenic steroid abusers are first exposed to the drugs before the age of 16.[9] Notable increases in AAS abuse have also been reported for adolescent girls.[7]

AAS were originally designed to promote maximal enhancement of protein synthesis and muscle growth while having minimal androgenic or masculinizing effects,[10] however all AAS have some androgenic effects.[11] AAS exert their effects by binding to the androgen receptor, which ultimately leads to an increase in protein synthesis.[12] AAS also antagonize glucocorticosteroid receptors and thereby prevent the catabolic effects of glucocorticoids.[13] Endogenous testosterone is not used for supplementation because it has a much lower potency than its derivatives and is more rapidly metabolized by the liver.[14] There are three main classes of AAS. The first class consists of compounds created by the esterification of the 17-β-hydroxyl group of testosterone and includes compounds such as testosterone propionate.[11] This class of AAS can be metabolized into three different compounds. They can either be hydrolyzed to free testosterone, reduced to 5-α-dihydroxytestosterone, which has a higher potency at androgen receptors than testosterone,[15,16] or aromatized to estrogens such as 17-β-estradiol. Therefore, class I metabolites can have activity at both androgen receptors and estrogen receptors. In addition, these neurosteroids may allosterically modulate $GABA_A$ and glutamate receptors.[17,18] The second class of AAS consists of 19-nortestosterone derivatives, such as nandrolone decanoate, which have a methyl group at C19.[1,10] Although to a lesser extent than testosterone, class II AAS can be aromatized to estrogens. The third class of AAS includes compounds that are alkylated at C-17. Examples of this group are 17α-methyltestosterone and stanozolol. This class of AAS are orally active as C-17-alkylation slows their metabolism by the liver.[1] Compounds in this class cannot be reduced to dihydroxytestosterone and therefore they have a lower potency at androgen receptors but fewer androgenic effects.[19] These compounds cannot be aromatized to 17-β-estradiol but can be converted to other androgenic and estrogenic compounds.[17,20]

The typical AAS abuser is a male polysubstance user with psychiatric comorbidity.[21-24] AAS abusers self-administer a variety of androgenic compounds orally, transdermally or by injection. As compared to the nanomolar concentrations of endogenous steroids found in serum, micromolar concentrations have been observed in human subjects who abuse AAS.[17] Stacking and pyramiding are common practices among AAS abusers. "Stacking" refers to the combined use of oral and injectable AAS for the purpose of increasing potency and balancing out side effects. Decreasing doses in weekly cycles is referred to as "pyramiding."[25] The abuse of AAS is associated with a large number of negative side effects.

ADVERSE EFFECTS OF AAS ABUSE

The abuse of AAS has been shown to result in numerous negative physiological and psychological consequences, some of which may be irreversible. These include reproductive effects, such as a decrease in the levels of luteinizing and follicle stimulating hormone, which leads to a decrease in testosterone production, decreased spermatogenesis, and testicular atrophy, all of which may be abnormal for up to six months after the discontinuation of steroid abuse.[26-28] In women, menstrual abnormalities and masculinizing effects are potentially irreversible.[29,30] Hepatic dysfunctions, including elevations in the levels of liver enzymes, are most commonly associated with the abuse of AAS from class III.[31,32] Both benign and malignant tumors of the liver have been detected.[31,33-35] AAS abuse has also been linked to irreversible changes in the myocardium such as concentric left ventricular hypertrophy[36] and possibly irreversible decreases in high-density lipoprotein, which may increase the risk for atherosclerotic heart disease.[10,37-39] AAS abuse among adolescents is particularly alarming because it has been shown to result in premature epiphyseal closure.[40] Endocrine effects such as altered glucose tolerance, increased insulin resistance, and decreases in thyroid hormones have also been reported.[10,32,41] The investigation of the negative health effects of AAS abuse is, however, complicated by several factors. It is difficult to discern what adverse effects are caused by a specific AAS because abusers are often unaware of what or how much AAS they are consuming.[42] In addition, AAS abuse is often associated with the abuse of other drugs such as cocaine, amphetamine, and heroin.[23,43]

With regard to the psychological changes associated with AAS abuse, a variety of adverse effects have been reported. Some studies have reported that individuals who abuse AAS experience an increase in irritability, mood swings, aggression, depression, altered libido, acute paranoia, delirium, mania, and even psychosis.[17,44-47]

AAS ABUSE AND NEGATIVE AFFECTIVE STATES

Studies that compared steroid abusing athletes with athletes who do not abuse steroids have shown that steroid abuse leads to a higher incidence and prevalence of psychiatric symptoms.[48-50] Variability in psychiatric symptoms among AAS abusers may be explained by differences in doses consumed, agents used, duration of use, personality type of the user, and concurrent or previous use of other recreational drugs.[51-54] Several clinical studies have investigated the effects of relatively high doses of testosterone (500-600 mg/week) on mental health.[55-59] It was shown that the majority of individuals do not exhibit psychiatric symptomology at the aforementioned doses. It remains unanswered whether these studies accurately model the abuse patterns of real-life AAS users when the majority of users not only consume weekly doses which greatly surpass 500 mg, but also ingest a variety of oral and injectable AAS simultaneously.[60,61] It has been suggested that psychotic symptoms associated with AAS use will only develop among individuals consuming more than 1,000 mg of testosterone per week.[50]

Su and colleagues investigated the acute effects of AAS on mood and behavior in healthy male volunteers.[57] An inpatient study was conducted, in which either a low or a high dose of methyltestosterone was administered over a two-week period. Subjects given the higher dose of 240 mg per day displayed small but significant changes in symptom scores on positive and negative mood scales as well as on scales of cognitive impairment as compared to baseline. An acute manic episode was reported by one of the twenty subjects. Neither family history nor history of previous drug use was a predictor of the development or the nature of psychiatric symptoms. Although the weekly dose administered in this study was high, the changes in psychological parameters were small. This might be explained by the administration of only one anabolic androgenic steroid, which does not parallel the real-life use patterns of abusers.

Recently, AAS abuse severity and its correlation to changes in psychiatric symptomology were evaluated in subjects following a self-directed regimen of self-obtained AAS.[62] Investigators compared the prevalence of psychopathological side effects in three groups: control, placebo, and AAS. AAS users in this study were administering 2 to 4 oral agents, in doses approximately 3 times higher than those used in controlled studies, in combination with 1 to 3 injectable agents, in doses 5 to 10 times greater than in controlled studies. The self-directed cycles implemented by anabolic androgenic steroid users lasted a mean of 9.4 weeks. Users were further subdivided into 3 groups (light, medium, and heavy abuse) based on the number, type, dose, and therapeutic index of agents used as well as the number and length of cycles. Psychiatric symptomology was evaluated before, midpoint and a week before the end of each cycle using the SCL-90, a 90 item self-report system inventory comprising nine dimensions of psychiatric symptomology. It was shown that AAS abuse induced a wide array of psychopathological side effects aside from aggression and hostility. Placebo effects were not observed. Not only did the psychological side effects escalate as the abuse patterns intensified, but at the re-evaluation, AAS users reported symptoms that were absent at the first evaluation. Particularly high scores were observed on the scales for obsessive compulsion, interpersonal sensitivity, anxiety, hostility, phobic anxiety, and paranoid ideation, with moderately high scores on the scales for somatization, depression, and psychoticism, among users. This was the first study to show that the extent of psychiatric symptoms is dependent upon the severity of AAS abuse, while monitoring subjects taking real-life doses and combinations of AAS, who were similar to placebo and control groups in all aspects aside from AAS use.

Other psychological disorders such as AAS dependency and withdrawal have also been reported among anabolic androgenic steroid users.[12,63]

ADDICTIVE PROPERTIES OF AAS

At this point in time, there is no evidence that AAS abuse or dependence can develop from the therapeutic use of AAS, however several studies have shown that chronic abuse may lead to dependence.[64-66] It has been reported that subjects who abuse AAS experience a state of euphoria[47,65,67] as well as increased aggression and diminished fatigue,[68] and that individuals who use high doses of AAS may develop depressive symptoms, anhedonia, fatigue, impaired concentration and even suicidal thoughts, when they stop taking AAS.[69] It appears that AAS, similar to other drugs of abuse, produce a wide spectrum of effects. The early stage of AAS abuse is characterized by euphoria while chronic abuse and discontinuation of abuse leads to a negative emotional state.

The endogenous secretion of testosterone is associated with positively reinforcing social behaviors such as mating and aggression.[70,71] Its secretion is stimulated following successful mating[72-75] and depressed during defeat.[74-76] Dominant animals that win more agonistic encounters and copulate with a higher frequency have higher levels of testosterone compared to subordinates.[73-76] This suggests that testosterone reinforces specific behavioral responses.[19]

In a two-stage model of AAS dependence as proposed by Brower and colleagues,[77] the anabolic effects on muscle growth account for the first stage of steroid use, which after prolonged exposure can lead to both physical and psychological dependence. During the last two decades, Brower and colleagues have conducted numerous studies that examined AAS dependence in humans.[64,65,77] They initially concluded that 75% of steroid users met the DSM-III-R criteria for psychoactive substance dependence and that all subjects met the criteria for psychoactive substance abuse. They later

observed a 60% rate of psychoactive dependence. They also concluded that withdrawal effects were more common for individuals using 17-α-alkylated AAS than for subjects using AAS from the other two classes.

Although it is evident that AAS are rewarding independent of their anabolic effects, defining the potential for addiction in humans has been proven difficult.[78] In humans, it is difficult to dissociate the psychological addiction–the dependence on AAS in order to maintain a desired muscle mass, from a physiological addiction, whose mechanisms may parallel those of other drugs of abuse.[79,80] Animal studies have contributed to a better understanding of the brain mechanism underlying AAS dependence. The administration of testosterone into the nucleus accumbens has been shown to induce conditioned place preference in rats.[81] This indicates that the administration of testosterone into parts of the brain reward system can induce a positive emotional state. However, both the ventral tegmental area, another component of the brain reward system, and the nucleus accumbens have few steroid receptors.[82] In addition, injections of testosterone directly into the medial preoptic area (a region rich in androgen receptors, and a key site for the organization of male sexual behavior) of rats failed to induce a conditioned place preference.[83] This supports the hypothesis that androgens may mediate their rewarding effects via the activation of non-steroid receptors. It has been shown that testosterone-induced conditioned place preference involves the release of dopamine in the nucleus accumbens,[81,84] which is a common mechanism shared by other drugs of abuse. In a more recent study,[85] it was observed that dopamine D1 and D2 receptor antagonists prevent testosterone-induced conditioned place preference. This suggests an important role for increased dopaminergic transmission in the rewarding effects of testosterone.

Although place preference studies show that androgens have rewarding properties that may play a role in the initiation of testosterone taking behavior, these studies do not provide evidence for the notion that chronic testosterone use could lead to compulsive testosterone taking behavior, and that the discontinuation of testosterone taking behavior is associated with a withdrawal syndrome (i.e., testosterone addiction). It has been shown that hamsters self-administer testosterone.[86,87] Recently Peters and Wood[78] reported that hamsters will self-administer testosterone to the point of death, which might be indicative of a loss of control over testosterone intake in a setting where psychological factors such as concern about athletic performance are irrelevant. It was observed that testosterone overdose resembles opioid overdose, in that its symptoms include reduced locomotion, decreased respiration and body temperature and that hamsters could develop both behavioral and physiological tolerance to these depressive symptoms following repeated infusions. In other studies, it was shown that AAS increase levels of endogenous opioids in the hypothalamus, striatum, and periaquaductal grey.[88,89] In addition, the opioid antagonist naltrexone blocks the reinforcing and depressive effects of testosterone.[78] This suggests that testosterone mediates its rewarding properties at least partly via activation of central opioid systems.

It has been suggested that androgen reinforcement is not comparable to that of cocaine or heroine, but is more similar to mild reinforcers such as caffeine, nicotine, or benzodiazepines.[19] However, symptoms of acute AAS withdrawal are similar to those observed during opioid withdrawal. Opioid withdrawal is characterized by anxiety, irritability, insomnia, hot flashes, sweats, chills, anorexia, myalgia, nausea, vomiting, piloerection, tachycardia, hypertension, depression, and craving.[22,63] It is therefore possible that AAS have a higher addiction liability then is presently believed.

MECHANISM OF AAS ACTION

Despite the wide array of responses mediated by different AAS, all compounds bind directly to one identified androgen receptor,[90] but have different binding affinities, which vary from tissue to tissue.[91] In males with normal physiologic levels of testosterone, the androgen receptors are saturated. Therefore, it has been suggested that there are other mechanisms by which AAS produce their effects.[1]

Allosteric Modulation of the GABA_A Receptor

In the basal forebrain, $GABA_A$ receptors play an important role in a variety of behaviors such as aggression, anxiety and reproductive behaviors. In both animals and humans, changes in these behaviors are the most prominent effects associated with anabolic androgenic steroid abuse.[11] Brain regions that could possibly mediate the expression of these behaviors are the medial preoptic area, ventromedial nucleus, and the medial amygdala.[17,18] The $GABA_A$ receptor is a pentameric inonotropic transmembrane protein for which 16 different receptor subunits genes ($\alpha 1$-$\alpha 6$, $\beta 1$-$\beta 3$, $\gamma 1$-$\gamma 3$, δ, ε, π, and θ) have been identified in mammals.[18] It has been shown that the type of α subunit determines the allosteric modulation by neurosteroids[92,93] and that the ε subunit is significant in that it is highly expressed in gonadotropin-releasing hormone neurons of the ventromedial nucleus and medial preoptic area.[94] GABAergic control of gonadotropin-releasing hormone pulsatility is essential for pubertal onset and estrous cycling.[95] In the basal forebrain, the expression of $GABA_A$ receptor subunit mRNA changes throughout development.[96] During all stages of development, receptors containing the $\alpha 2$ subunit predominate in forebrain regions, while in the adult brain the most common $GABA_A$ receptor isoform is composed of $\alpha 1$, $\beta 2$, or $\beta 3$ and $\gamma 2$ subunits.[17,18] In adult animals, it has been shown that $GABA_A$ receptor expression and function are dependent upon hormonal state.[96] In addition, it has been shown that in mice, following a 4 week exposure to 17-α-methyl-4-androsten-17-β-ol-3-one (17αmet), there were significant decreases in the levels of $\alpha_{1,2,5}$, $\gamma_{1,2}$ and ε mRNA. The most significant changes were observed in the levels of α subunit mRNA in female mice. With regard to changes in GABAergic systems, it appears that female mice and in particular adolescent female mice are the most sensitive to anabolic androgenic steroid exposure.[11] It is possible that the observed changes in $GABA_A$ receptor subunit composition that result from chronic exposure to AAS may alter the sensitivity of the central nervous system to the actions of AAS as well as to other endogenous and exogenous agents.[11] Of particular interest is the impact of AAS on the developing GABAergic system, since adolescent AAS use has been increasing in recent years,[7] and to what extent these changes could be reversible or permanently alter the course of development.

It is hypothesized that the acute effects of AAS, which include decreases in anxiety and an enhanced sense of well being are at least in part the result of allosteric enhancement of GABAergic transmission, while the effects of chronic exposure, such as increased aggression, anxiety and inhibition of reproductive behaviors, may result from the down-regulation of $GABA_A$ receptors.[11]

Neuropeptides

Substance P (SP) is an excitatory neurotransmitter/neuromodulator that is part of the tachykinin family of peptides and is widely distributed in the mammalian CNS as well as peripheral tissues.[97] It is implicated in memory processing, motor activities, reinforcement, endocrine functions, and nociception.[98,99] It is also believed to play an important role as a modulator in the expression of anxiety, depression, and aggression,[100,101] which are prominent symptoms of AAS abuse. Following administration of nandrolone decanoate, the levels of substance P have been shown to be significantly increased in several brain regions, including the hypothalamus, amygdala, periaquaductal grey and striatum.[102]

The actions of SP are terminated due to its degradation by substance P endopeptidase, which generates biologically active fragments.[103,104] Levels of the N-terminal fragment of substance P (SP1-7), which is believed to exert similar as well as opposing effects,[103] are also altered, following anabolic androgenic steroid administration.[102] The increases in substance P resulting from AAS treatment may be explained by decreases in the activity of substance P endopeptidase, which has been observed in the hypothalamus, substantia nigra, caudate putamen, and ventral tegmental area following intramuscular injections of nandrolone in rats.[105]

Although substance P is up-regulated as a result of AAS administration, in another study it was shown that the levels of NK1, the major neurokinin receptor for substance P are down-regulated in rats, following 2 weeks of intramuscular injections of nandrolone in brain regions associated with aggression, anxiety, fear processing, impulsivity and memory function.[106] The density of NK1 receptors was reduced in the medial hypothalamus, particularly in the dorsomedial hypothalamic nucleus, an area implicated in aggressive and defensive behavior.[107] Significant decreases in the basolateral amygdala, involved in memory, decision-making and in adaptive response selection to affective stimuli,[108-110] occurred. A significant down-regulation was also observed in the nucleus accumbens core. The decrease in NK1 receptors in the periaquaductal grey is also significant, in that it may explain the increase in anxiety, which has been reported by individuals who abuse AAS. The midbrain periaquaductal grey is a major site of the analgesic actions of opioids.[111] Noxious stimuli and stress induce the release of substance P within the periaquaductal grey, which produces analgesia.[112,113] It was concluded that the down-regulation of the NK1 receptors in the aforementioned regions may be the result of feedback regulation mechanisms such as receptor internalization.[106] NK1 receptor antagonists may prove useful in the treatment of AAS dependence and withdrawal symptoms, since clinical studies have suggested their potential therapeutic use in the treatment of affective disorders such as depression.[114,115] The antidepressive effects of NK1 receptor antagonism have been attributed to enhanced serotonergic neurotransmission.[116]

Hypothalamic-Pituitary-Adrenal Axis

Several lines of evidence indicate that chronic administration of AAS results in a disruption of the regulation of the Hypothalamic-Pituitary-Adrenal (HPA) axis. High doses of AAS have been reported to alter the expression of proopiomelanocortin (POMC) and corticotropin-releasing factor (CRF) mRNA in rat brain and the circulating levels of corticosterone and adrenocorticotropin hormone (ACTH) of rats.[117] It has also been suggested that the overactivity of central CRF systems is involved in the negative affective state associated with chronic drug abuse and withdrawal[118] and that CRF in the amygdala, plays an important role in mediating psychiatric symptoms similar to those associated with AAS abuse and withdrawal.[119,120]

The acute effects of high dose nandrolone decanoate on the HPA-axis of rats have been assessed.[121] In order to determine the temporal nature of the changes in HPA-axis hormones, rats were sacrificed either 1 hour or 24 hours after the last nandrolone injection. Circulating levels of ACTH and corticosterone were increased 1 but not 24 hours after the last injection as compared to controls. In addition, levels of hypothalamic proopiomelanocortin and amygdaloid CRF mRNA were significantly lower 24 hours as compared to 1 hour after the last injection.

Conversely, in humans, it was observed that acute methyltestosterone administration does not produce significant changes in the levels of HPA hormones.[122] On the other hand, chronic (two weeks) administration of nandrolone decanoate has been shown to decrease plasma ACTH and corticosterone levels compared to baseline levels in rats.[117] The present findings indicate that AAS have an effect on the HPA-axis, but further research on the acute and chronic effects is needed before firm conclusions can be drawn.

In a recent study, conducted by Hamke and colleagues,[123] it was demonstrated that the activation of the NK1 receptor by substance P induces the expression of cotricotropin-releasing factor receptor 1 (CRF1) in a human astrocytoma cell line as well as in primary rat astroglial cells. On the other hand, in healthy humans, intravenous substance P increases circulating levels of both ACTH and cortisol.[124,125] These studies suggest a mechanism which links the up-regulated SP and effects on the HPA axis during AAS use. The increased levels of SP result in the increased expression of CRF1 receptors and perhaps to anxiogenic effects, which are characteristic of chronic AAS use. These studies also provide further evidence for the possible use of NK1 antagonists in the treatment of anxiety and depression associated with AAS. More research is however necessary in order to elucidate the effects of AAS on the HPA axis as well as its modulation by SP.

Monoaminergic Systems

Preclinical studies suggest that AAS abuse may lead to an increase in aggressive behavior, which may be partly mediated by the effects of AAS on serotonergic transmission.[126,127] A connection between AAS and central dopaminergic activity has also been established in animal studies.[43,128] Within the dorsal raphe nuclei, numerous 5-HT neurons are co-localized with non-5-HT neurons, including GABA, glutamate and dopamine (DA) positive neurons, that can regulate the activity of 5-HT neurons through local circuits or mediate the effects of afferent inputs.[129] Other mediators of the dorsal raphe include CRF and substance P. In a recent immunolabeling study,[130] it was demonstrated that CRF axon terminals contact GABA immunolabeled dendrites in the dorsal raphe, which indicates that GABA may act as an intermediate between CRF and 5-HT.

Kurling and colleagues[138] investigated the effects of nandrolone decanoate on the levels of brain dopamine and serotonin. They showed that sub-chronic treatment with supraphysiological doses of nandrolone decanoate induces changes in both the dopaminergic and serotonergic transmission in rats. Levels of DOPAC, a metabolite of DA, as well as levels of 5-HT were increased in the cerebral cortex. An up-regulation of monoamine oxidase (MAO)-A, which metabolizes DA to DOPAC and, could be a possible explanation for the higher concentrations of DOPAC.[138] An up-regulation of the DA transporter, which also increases DA turnover to DOPAC, could also explain the increases in DOPAC levels. In a previous study, increased levels of DA transporter proteins on presynaptic terminals following chronic nandrolone treatment were reported.[139,140] This group was unable to determine whether the increases in DOPAC reflected increases in biosynthesis, secretion or both, however it was speculated that the subchronic treatment up-regulated biosynthesis.[138] Further investigation into the effects of AAS on the mesolimbic dopaminergic system are warranted as this system plays an important role in brain reward function.[141] The activation of this system, especially during the early stages of AAS abuse, could explain the reported feelings of euphoria, increased self-esteem and confidence. Chronic treatment with nandrolone also alters serotonin levels in several brain regions. Increased 5-HT concentrations were observed in the cerebral cortex, hypothalamus and hippocampus.[138] Increases in 5-HT transmission have been linked to a reduction in aggressive behavior.[142-144] Since aggression is more often linked to prolonged AAS use, it is likely that secondary adaptations occur within serotonergic circuits after prolonged use.[138] Since the dorsal raphe nucleus is so highly regulated by other neurotransmitter/neuromodulator systems, which have also been shown to be affected by AAS, it is possible that changes in these systems alter serotonergic circuits during chronic use, which results in increased aggression. A likely candidate would be the GABAergic projections, since it has been shown that AAS have opposing acute and chronic effects on GABA.[11]

CONCLUSIONS

The acute phase of anabolic AAS use is typically characterized by enhanced energy, self-esteem, self-confidence, euphoria and libido as well as irritability, anger and aggression,[57,145] while chronic abuse is associated with symptoms such as depression, anxiety, suspiciousness, disinhibition and other physiological effects.[50,146] In order to begin to answer the question of what mechanisms are involved in enabling the switch from euphoric to dysphoric symptomology during chronic AAS use, one has to consider the numerous brain regions and neurotransmitter systems which are modulated by AAS and their metabolites. Neurotransmitter/neuromodulator systems affected by AAS include the opioid system, in that levels of opioid peptides are altered in rats following anabolic androgenic steroid treatment.[88,89] Chronic AAS treatment has been shown to alter the numbers of both dopaminergic and serotonergic receptors in the rat brain.[139,147] The glutamatergic[148,149] and GABAergic[96] systems are also affected. The modulation of substance P and CRF should be further investigated since these peptides are believed to play an important role in affective disorders such as anxiety and depression. Taken together, these studies suggest that AAS, like other drugs of abuse, mediate their rewarding effects via the activation of brain dopamine, serotonin, and opioid systems. The activation of brain CRF systems could possibly play a role in the negative emotional state associated with chronic AAS abuse.

REFERENCES

1. Basaria S, Wahlstrom JT, Dobs AS. Clinical review 138: Anabolic-androgenic steroid therapy in the treatment of chronic diseases. J Clin Endocrinol Metab. 2001; 86:5108-17.

2. Brower KJ. Anabolic steroid abuse and dependence. Curr Psychiatry Rep. 2002; 4:377-87.

3. Rabkin JG, Ferrando SJ, Wagner GJ, Rabkin R. DHEA treatment for HIV + patients: effects on mood, androgenic and anabolic parameters. Psychoneuroendocrinology. 2000; 25:53-68.

4. Friedl KE. Reappraisal of the health risks associated with the use of high doses of oral and injectable androgenic steroids. NIDA Res Monogr. 1990; 102: 142-77.

5. Kopera H. Interactions of anabolic steroids. Wien Med Wochenschr. 1993; 143:401-2.

6. Goldstein PJ. Anabolic steroids: an ethnographic approach. NIDA Res Monogr. 1990; 102:74-96.

7. Johnston LD, O'Malley PM, Bachman JG, Schulenberg JE. Secondary school students. National Institutes on Drug Abuse. Bethesda, MD: NIH Publication No. 04-5507. 2004; 545.

8. Yesalis CE, Anderson WA, Buckley WE, Wright JE. Incidence of the nonmedical use of anabolic-androgenic steroids. NIDA Res Monogr. 1990; 102:97-112.

9. Buckley WE, Yesalis CE, III, Friedl KE, Anderson WA, Streit AL, Wright JE. Estimated prevalence of anabolic steroid use among male high school seniors. JAMA. 1988; 260:3441-5.

10. Shahidi NT. A review of the chemistry, biological action, and clinical applications of anabolic-androgenic steroids. Clin Ther. 2001; 23:1355-90.

11. Henderson LP, Penatti CA, Jones BL, Yang P, Clark AS. Anabolic androgenic steroids and forebrain GABAergic transmission. Neuroscience. 2006; 138:793-9.

12. Bahrke MS, Yesalis CE. Abuse of anabolic androgenic steroids and related substances in sport and exercise. Curr Opin Pharmacol. 2004; 4:614-20.

13. Haupt HA, Rovere GD. Anabolic steroids: a review of the literature. Am J Sports Med. 1984; 12: 469-84.

14. Sturmi JE, Diorio DJ. Anabolic agents. Clin Sports Med. 1998; 17:261-82.

15. Winters SJ. Androgens: endocrine physiology and pharmacology. NIDA Res Monogr. 1990; 102:113-30.

16. Kochakian CD. History, chemistry and pharmacodynamics of anabolic-androgenic steroids. Wien Med Wochenschr. 1993; 143:359-63.

17. Clark AS, Henderson LP. Behavioral and physiological responses to anabolic-androgenic steroids. Neurosci Biobehav Rev. 2003; 27:413-36.

18. Henderson LP, Jorge JC. Steroid modulation of GABAa receptors. CNS roles in reproduction, dysfunction and drug abuse. In: Maue RA (Ed.), Advances in Molecular and Cell Biology: Molecular Insights into Ion Channel Biology in Health and Disease. Amsterdam, Elsevier, 2004:217-249.

19. Wood RI. Reinforcing aspects of androgens. Physiol Behav. 2004; 83:279-89.

20. Kammerer RC. Drug testing and anabolic steroids. In: Yesalis CE (Ed.), Anabolic Steroids in Sports and Exercise. 2 ed. Champaign, IL: Human Kinetics, 2000:415-60.

21. Yesalis CE, Wright JE, Lombardo JA. Anabolic steroids in athletes. Wien Med Wochenschr. 1992; 142: 298-308.

22. Bahrke MS, Yesalis CE, Kopstein AN, Stephens JA. Risk factors associated with anabolic-androgenic steroid use among adolescents. Sports Med. 2000; 29:397-405.

23. DuRant RH, Escobedo LG, Heath GW. Anabolic-steroid use, strength training, and multiple drug use among adolescents in the United States. Pediatrics. 1995; 96:23-8.

24. Perry PJ, Kutscher EC, Lund BC, Yates WR, Holman TL, Demers L. Measures of aggression and mood changes in male weightlifters with and without androgenic anabolic steroid use. J Forensic Sci. 2003; 48:646-51.

25. Karch SB. Karch's pathology of drug abuse. 3 ed. Boca Raton: CRC Press, 2002.

26. Boyadjiev NP, Georgieva KN, Massaldjieva RI, Gueorguiev SI. Reversible hypogonadism and azoospermia as a result of anabolic-androgenic steroid use in a bodybuilder with personality disorder. A case report. J Sports Med Phys Fitness. 2000; 40:271-4.

27. Dohle GR, Smit M, Weber RF. Androgens and male fertility. World J Urol. 2003; 21:341-5.

28. Eklof AC, Thurelius AM, Garle M, Rane A, Sjoqvist F. The anti-doping hot-line, a means to capture the abuse of doping agents in the Swedish society and a new service function in clinical pharmacology. Eur J Clin Pharmacol. 2003; 59:571-7.

29. American Academy of Pediatrics. Adolescents and anabolic steroids: a subject review. Pediatrics. 1997; 99:904-8.

30. Elliot DL, Goldberg L. Women and anabolic steroids. In: Yesalis CE (Ed.), Anabolic Steroids in Sports Medicine. 2 ed. Champaign: Human Kinetics, 2000: 225-46.

31. Friedl KE. Effect of Anabolic steroids on physical health. In: Yesalis CE (Ed.), Anabolic Steroids in Sport and Exercise. 2 ed. Champaign: Human Kinetics, 2000:175-225.

32. Snyder PJ. Androgens. In: Hardman JG, Limbird LE, Goodman Gilman A (Eds.), The Pharmacological Basis of Therapeutics. 10 ed. New York: McGraw Hill, 2001:1635-48.

33. Watanabe S, Kobayashi Y. Exogenous hormones and human cancer. Jpn J Clin Oncol. 1993; 23:1-13.

34. Soe KL, Soe M, Gluud CN. Liver pathology associated with anabolic androgenic steroids. Ugeskr Laeger. 1994; 156:2585-8.

35. Velazquez I, Alter BP. Androgens and liver tumors: Fanconi's anemia and non-Fanconi's conditions. Am J Hematol. 2004; 77:257-67.

36. Urhausen A, Albers T, Kindermann W. Are the cardiac effects of anabolic steroid abuse in strength athletes reversible? Heart. 2004; 90:496-501.

37. Hartgens F, Rietjens G, Keizer HA, Kuipers H, Wolffenbuttel BH. Effects of androgenic-anabolic steroids on apolipoproteins and lipoprotein (a). Br J Sports Med. 2004; 38:253-9.

38. Glazer G. Atherogenic effects of anabolic steroids on serum lipid levels. A literature review. Arch Intern Med. 1991; 151:1925-33.

39. Cohen JC, Noakes TD, Benade AJ. Hypercholesterolemia in male power lifters using anabolic-androgenic steroids. Physician Sportsmedecine. 1988; 16:49-56.

40. Al Ismail K, Torreggiani WC, Munk PL, Nicolaou S. Gluteal mass in a bodybuilder: radiological depiction of a complication of anabolic steroid use. Eur Radiol. 2002; 12:1366-9.

41. Cohen JC, Hickman R. Insulin resistance and diminished glucose tolerance in powerlifters ingesting anabolic steroids. J Clin Endocrinol Metab. 1987; 64:960-3.

42. Medical aspects of drug use in the gym. Drug Ther Bull. 2004; 42:1-5.

43. Kindlundh AM, Hagekull B, Isacson DG, Nyberg F. Adolescent use of anabolic-androgenic steroids and relations to self-reports of social, personality and health aspects. Eur J Public Health. 2001; 11:322-8.

44. Wilson IC, Prange AJ, Jr., Lara PP. Methyltestosterone with imipramine in men: conversion of depression to paranoid reaction. Am J Psychiatry. 1974; 131:21-4.

45. Bahrke MS, Wright JE, O'Connor JS, Strauss RH, Catlin DH. Selected psychological characteristics of anabolic-androgenic steroid users. N Engl J Med. 1990; 323:834-5.

46. Middleman AB, DuRant RH. Anabolic steroid use and associated health risk behaviours. Sports Med. 1996; 21:251-5.

47. Kashkin KB, Kleber HD. Hooked on hormones? An anabolic steroid addiction hypothesis. JAMA. 1989; 262:3166-70.

48. Choi P, Parrott A, Cowan D. High-dose anabolic steroids in strength athletes: effects upon hostility and aggression. Hum Psychopharm. 1990; 5:349-56.

49. Lefavi R, Reeve T, Newland MC. Relationship between anabolic steroid use and selected psychological parameters in male bodybuilders. J Sport Behav. 1990; 13:157-66.

50. Pope HG, Jr., Katz DL. Psychiatric and medical effects of anabolic-androgenic steroid use. A controlled

study of 160 athletes. Arch Gen Psychiatry. 1994; 51: 375-82.

51. Hall RC, Popkin MK, Stickney SK, Gardner ER. Presentation of the steroid psychoses. J Nerv Ment Dis. 1979; 167:229-36.

52. Hall RC, Reading A. Psychosocial factors in medicine: steroid psychosis. New Physician. 1971; 20: 255-8.

53. Hall RC. Psychiatric adverse drug reactions: steroid psychosis. Clin Advances Treatm Psychiatr Disord. 1991:8-10.

54. Porcerelli JH, Sandler BA. Anabolic-androgenic steroid abuse and psychopathology. Psychiatr Clin North Am. 1998; 21:829-33.

55. Kouri EM, Lukas SE, Pope HG, Jr., Oliva PS. Increased aggressive responding in male volunteers following the administration of gradually increasing doses of testosterone cypionate. Drug Alcohol Depend. 1995; 40:73-9.

56. Pope HG, Jr., Kouri EM, Hudson JI. Effects of supraphysiologic doses of testosterone on mood and aggression in normal men: a randomized controlled trial. Arch Gen Psychiatry. 2000; 57:133-40.

57. Su TP, Pagliaro M, Schmidt PJ, Pickar D, Wolkowitz O, Rubinow DR. Neuropsychiatric effects of anabolic steroids in male normal volunteers. JAMA. 1993; 269:2760-4.

58. Tricker R, Casaburi R, Storer TW et al. The effects of supraphysiological doses of testosterone on angry behavior in healthy eugonadal men–a clinical research center study. J Clin Endocrinol Metab. 1996; 81:3754-8.

59. Yates WR, Perry PJ, MacIndoe J, Holman T, Ellingrod V. Psychosexual effects of three doses of testosterone cycling in normal men. Biol Psychiatry. 1999; 45:254-60.

60. Inigo MA, Arrimadas E, Arroyo D. 43 cycles of anabolic steroid treatment studied in athletes: the uses and secondary effects. Rev Clin Esp. 2000; 200:133-8.

61. Fudala PJ, Weinrieb RM, Calarco JS, Kampman KM, Boardman C. An evaluation of anabolic-androgenic steroid abusers over a period of 1 year: seven case studies. Ann Clin Psychiatry. 2003; 15:121-30.

62. Pagonis TA, Angelopoulos NV, Koukoulis GN, Hadjichristodoulou CS. Psychiatric side effects induced by supraphysiological doses of combinations of anabolic steroids correlate to the severity of abuse. Eur Psychiatry. 2006; 21:551-562.

63. Foley JD, Schydlower M. Anabolic steroid and ergogenic drug use by adolescents. Adolesc Med. 1993; 4:341-52.

64. Brower KJ, Blow FC, Beresford TP, Fuelling C. Anabolic-androgenic steroid dependence. J Clin Psychiatry. 1989; 50:31-3.

65. Brower KJ, Blow FC, Young JP, Hill EM. Symptoms and correlates of anabolic-androgenic steroid dependence. Br J Addict. 1991; 86:759-68.

66. Clark AS, Lindenfeld RC, Gibbons CH. Anabolic-androgenic steroids and brain reward. Pharmacol Biochem Behav. 1996; 53:741-5.

67. Cicero TJ, O'Connor LH. Abuse liability of anabolic steroids and their possible role in the abuse of alcohol, morphine, and other substances. NIDA Res Monogr. 1990; 102:1-28.

68. Wilson JD, Griffin JE. The use and misuse of androgens. Metabolism 1980; 29:1278-95.

69. Hall RC, Hall RC, Chapman MJ. Psychiatric complications of anabolic steroid abuse. Psychosomatics. 2005; 46:285-90.

70. Siegel HI. Male sexual behavior. In: Siegel HI (Ed.), The Hamster: Reproduction and Behavior. New York: Plenum, 1985:191-206.

71. Siegel HI. Aggressive behavior. In: Siegel HI (Ed.), The Hamster: Reproduction and Behavior. New York: Plenum, 1985:261-88.

72. Monaghan EP, Glickman SE. Hormones and aggressive behavior. In: Becker J, Breedlove M, Crews D (Eds.), Behavioral Endocrinology. Cambridge: MIT Press, 1992:261-86.

73. Macrides F, Bartke A, Fernandez F, D'Angelo W. Effects of exposure to vaginal odor and receptive females on plasma testosterone in the male hamster. Neuroendocrinology. 1974; 15:355-64.

74. Mazur A, Booth A. Testosterone and dominance in men. Behav Brain Sci. 1998; 21:353-63.

75. Niikura S, Yokoyama O, Komatsu K, Yotsuyanagi S, Mizuno T, Namiki M. A causative factor of copulatory disorder in rats following social stress. J Urol. 2002; 168:843-9.

76. Zhang JX, Zhang ZB, Wang ZW. Scent, social status, and reproductive condition in rat-like hamsters (Cricetulus triton). Physiol Behav. 2001; 74:415-20.

77. Brower KJ, Eliopulos GA, Blow FC, Catlin DH, Beresford TP. Evidence for physical and psychological dependence on anabolic androgenic steroids in eight weight lifters. Am J Psychiatry. 1990; 147:510-2.

78. Peters KD, Wood RI. Androgen dependence in hamsters: overdose, tolerance, and potential opioidergic mechanisms. Neuroscience. 2005; 130:971-81.

79. Frye CA, Seliga AM. Testosterone increases analgesia, anxiolysis, and cognitive performance of male rats. Cogn Affect Behav Neurosci. 2001; 1:371-81.

80. Aikey JL, Nyby JG, Anmuth DM, James PJ. Testosterone rapidly reduces anxiety in male house mice (Mus musculus). Horm Behav. 2002; 42:448-60.

81. Packard MG, Cornell AH, Alexander GM. Rewarding affective properties of intra-nucleus accumbens injections of testosterone. Behav Neurosci. 1997; 111: 219-24.

82. Kritzer MF. Selective colocalization of immunoreactivity for intracellular gonadal hormone receptors and tyrosine hydroxylase in the ventral tegmental area, substantia nigra, and retrorubral fields in the rat. J Comp Neurol. 1997; 379:247-60.

83. King BE, Packard MG, Alexander GM. Affective properties of intra-medial preoptic area injections of testosterone in male rats. Neurosci Lett. 1999; 269: 149-52.

84. Packard MG, Schroeder JP, Alexander GM. Expression of testosterone conditioned place preference is blocked by peripheral or intra-accumbens injection of alpha-flupenthixol. Horm Behav. 1998; 34:39-47.

85. Schroeder JP, Packard MG. Role of dopamine receptor subtypes in the acquisition of a testosterone conditioned place preference in rats. Neurosci Lett. 2000; 282:17-20.

86. Johnson LR, Wood RI. Oral testosterone self-administration in male hamsters. Neuroendocrinology. 2001; 73:285-92.

87. Wood RI, Johnson LR, Chu L, Schad C, Self DW. Testosterone reinforcement: intravenous and intra-cerebroventricular self-administration in male rats and hamsters. Psychopharmacology (Berl). 2004; 171:298-305.

88. Johansson P, Ray A, Zhou Q, Huang W, Karlsson K, Nyberg F. Anabolic androgenic steroids increase beta-endorphin levels in the ventral tegmental area in the male rat brain. Neurosci Res. 1997; 27:185-9.

89. Johansson P, Hallberg M, Kindlundh A, Nyberg F. The effect on opioid peptides in the rat brain, after chronic treatment with the anabolic androgenic steroid, nandrolone decanoate. Brain Res Bull. 2000; 51:413-8.

90. Feldkoren BI, Andersson S. Anabolic-androgenic steroid interaction with rat androgen receptor in vivo and in vitro: a comparative study. J Steroid Biochem Mol Biol. 2005; 94:481-7.

91. Karila T, Hovatta O, Seppala T. Concomitant abuse of anabolic androgenic steroids and human chorionic gonadotrophin impairs spermatogenesis in power athletes. Int J Sports Med. 2004; 25:257-63.

92. Yang P, Jones BL, Henderson LP. Role of the alpha subunit in the modulation of GABA(A) receptors by anabolic androgenic steroids. Neuropharmacology 2005; 49:300-16.

93. Yang P, Jones BL, Henderson LP. Mechanisms of anabolic androgenic steroid modulation of alpha(1)beta(3)gamma(2L) GABA(A) receptors. Neuropharmacology. 2002; 43:619-33.

94. Moragues N, Ciofi P, Lafon P, Tramu G, Garret M. GABAA receptor epsilon subunit expression in identified peptidergic neurons of the rat hypothalamus. Brain Res. 2003; 967:285-9.

95. Ojeda SR, Urbanski HF. Puberty in the rat. In: Knobil E, Neill JD (Eds.), The Physiology of Reproduction. New York: Raven Press Ltd., 1994:363-410.

96. McIntyre KL, Porter DM, Henderson LP. Anabolic androgenic steroids induce age-, sex-, and dose-dependent changes in GABA(A) receptor subunit mRNAs in the mouse forebrain. Neuropharmacology. 2002; 43:634-45.

97. Otsuka M, Yoshioka K. Neurotransmitter functions of mammalian tachykinins. Physiol Rev. 1993; 73:229-308.

98. Oitzl MS, Hasenohrl RU, Huston JP. Reinforcing effects of peripherally administered substance P and its C-terminal sequence pGlu6-SP6-11 in the rat. Psychopharmacology (Berl). 1990; 100:308-15.

99. Krappmann P, Hasenohrl RU, Frisch C, Huston JP. Self-administration of neurokinin substance P into the ventromedial caudate-putamen in rats. Neuroscience. 1994; 62:1093-101.

100. De Felipe C, Herrero JF, O'Brien JA et al. Altered nociception, analgesia and aggression in mice lacking the receptor for substance P. Nature. 1998; 392:394-7.

101. Gadd CA, Murtra P, De Felipe C, Hunt SP. Neurokinin-1 receptor-expressing neurons in the amygdala modulate morphine reward and anxiety behaviors in the mouse. J Neurosci. 2003; 23:8271-80.

102. Hallberg M, Johansson P, Kindlundh AM, Nyberg F. Anabolic-androgenic steroids affect the content of substance P and substance P(1-7) in the rat brain. Peptides. 2000; 21:845-52.

103. Hallberg M, Nyberg F. Neuropeptide conversion to bioactive fragments—an important pathway in neuromodulation. Curr Protein Pept Sci. 2003; 4:31-44.

104. Persson S, Le Greves P, Thornwall M, Eriksson U, Silberring J, Nyberg F. Neuropeptide converting and processing enzymes in the spinal cord and cerebrospinal fluid. Prog Brain Res. 1995; 104:111-30.

105. Magnusson K, Hallberg M, Hogberg AM, Nyberg F. Administration of the anabolic androgenic steroid nandrolone decanoate affects substance P endopeptidase-like activity in the rat brain. Peptides. 2006; 27:114-21.

106. Hallberg M, Kindlundh AM, Nyberg F. The impact of chronic nandrolone decanoate administration on the NK1 receptor density in rat brain as determined by autoradiography. Peptides. 2005; 26:1228-34.

107. Canteras NS. The medial hypothalamic defensive system: hodological organization and functional implications. Pharmacol Biochem Behav. 2002; 71:481-91.

108. McGaugh JL, Cahill L, Roozendaal B. Involvement of the amygdala in memory storage: interaction with other brain systems. Proc Natl Acad Sci U S A. 1996; 93:13508-14.

109. Baxter MG, Parker A, Lindner CC, Izquierdo AD, Murray EA. Control of response selection by reinforcer value requires interaction of amygdala and orbital prefrontal cortex. J Neurosci. 2000; 20:4311-9.

110. Killcross S, Robbins TW, Everitt BJ. Different types of fear-conditioned behaviour mediated by separate nuclei within amygdala. Nature. 1997; 388:377-80.

111. Yaksh TL, Yeung JC, Rudy TA. Systematic examination in the rat of brain sites sensitive to the direct application of morphine: observation of differential effects within the periaqueductal gray. Brain Res. 1976; 114:83-103.

112. Malick JB, Goldstein JM. Analgesic activity of substance P following intracerebral administration in rats. Life Sci. 1978; 23:835-44.

113. Rosen A, Brodin K, Eneroth P, Brodin E. Short-term restraint stress and s.c. saline injection alter the tissue levels of substance P and cholecystokinin in the peri-aqueductal grey and limbic regions of rat brain. Acta Physiol Scand. 1992; 146:341-8.

114. Kramer MS, Cutler N, Feighner J et al. Distinct mechanism for antidepressant activity by blockade of central substance P receptors. Science. 1998; 281:1640-5.

115. Kramer MS, Winokur A, Kelsey J et al. Demonstration of the efficacy and safety of a novel substance P (NK1) receptor antagonist in major depression. Neuropsychopharmacology. 2004; 29:385-92.

116. Adell A. Antidepressant properties of substance P antagonists: relationship to monoaminergic mechanisms? Curr Drug Targets CNS Neurol Disord. 2004; 3:113-21.

117. Zhou Y, Johansson P, Ho A, Nyberg F, Kreek MJ. Anabolic Androgenic Steroids Alter POMC mRNA Expression and HPA Activity in Male Rats. Problems of Drug Dependence, 1998: Proceedings of the 60th Annual Scientific Meeting of the College on Problems of Drug Dependency; Washington, DC: National Institute of Drug Abuse Research Monograph Series, 1998: 115.

118. Bruijnzeel AW, Gold MS. The role of corticotropin-releasing factor-like peptides in cannabis, nicotine, and alcohol dependence. Brain Res Rev. 2005; 49:505-28.

119. Kreek MJ, Koob GF. Drug dependence: stress and dysregulation of brain reward pathways. Drug Alcohol Depend. 1998; 51:23-47.

120. Koob GF, Le Moal M. Drug abuse: hedonic homeostatic dysregulation. Science. 1997; 278:52-8.

121. Schlussman SD, Zhou Y, Johansson P et al. Effects of the androgenic anabolic steroid, nandrolone decanoate, on adrenocorticotropin hormone, corticosterone and proopiomelanocortin, corticotropin releasing factor (CRF) and CRF receptor1 mRNA levels in the hypothalamus, pituitary and amygdala of the rat. Neurosci Lett. 2000; 284:190-4.

122. Daly RC, Su TP, Schmidt PJ, Pagliaro M, Pickar D, Rubinow DR. Neuroendocrine and behavioral effects of high-dose anabolic steroid administration in male normal volunteers. Psychoneuroendocrinology. 2003; 28:317-31.

123. Hamke M, Herpfer I, Lieb K, Wandelt C, Fiebich BL. Substance P induces expression of the corticotropin-releasing factor receptor 1 by activation of the neurokinin-1 receptor. Brain Res. 2006; 1102:135-44.

124. Coiro V, Capretti L, Volpi R et al. Stimulation of ACTH/cortisol by intravenously infused substance P in normal men: inhibition by sodium valproate. Neuroendocrinology. 1992; 56:459-63.

125. Lieb K, Ahlvers K, Dancker K et al. Effects of the neuropeptide substance P on sleep, mood, and neuroendocrine measures in healthy young men. Neuropsychopharmacology. 2002; 27:1041-9.

126. Miczek KA, Weerts E, Haney M, Tidey J. Neurobiological mechanisms controlling aggression: preclinical developments for pharmacotherapeutic interventions. Neurosci Biobehav Rev. 1994; 18:97-110.

127. Lindqvist AS, Johansson-Steensland P, Nyberg F, Fahlke C. Anabolic androgenic steroid affects competitive behaviour, behavioural response to ethanol and brain serotonin levels. Behav Brain Res. 2002; 133: 21-9.

128. Thiblin I, Finn A, Ross SB, Stenfors C. Increased dopaminergic and 5-hydroxytryptaminergic activities in male rat brain following long-term treatment with anabolic androgenic steroids. Br J Pharmacol. 1999; 126: 1301-6.

129. Valentino RJ, Commons KG. Peptides that fine-tune the serotonin system. Neuropeptides. 2005; 39:1-8.

130. Waselus M, Valentino RJ, Van Bockstaele EJ. Ultrastructural evidence for a role of gamma-aminobutyric acid in mediating the effects of corticotropin-releasing factor on the rat dorsal raphe serotonin system. J Comp Neurol. 2005; 482:155-65.

131. Nemeroff CB. The corticotropin-releasing factor (CRF) hypothesis of depression: new findings and new directions. Mol Psychiatry. 1996; 1:336-42.

132. Holsboer F. The rationale for corticotropin-releasing hormone receptor (CRH-R) antagonists to treat depression and anxiety. J Psychiatr Res. 1999; 33:181-214.

133. Gold PW, Chrousos GP. Organization of the stress system and its dysregulation in melancholic and atypical depression: high vs low CRH/NE states. Mol Psychiatry. 2002; 7:254-75.

134. Neckers LM, Schwartz JP, Wyatt RJ, Speciale SG. Substance P afferents from the habenula innervate the dorsal raphe nucleus. Exp Brain Res. 1979; 37: 619-23.

135. Li YQ, Rao ZR, Shi JW. Substance P-like immunoreactive neurons in the nucleus tractus solitarii of the rat send their axons to the nucleus accumbens. Neurosci Lett. 1990; 120:194-6.

136. Gradin K, Qadri F, Nomikos GG, Hillegaart V, Svensson TH. Substance P injection into the dorsal raphe increases blood pressure and serotonin release in hippocampus of conscious rats. Eur J Pharmacol. 1992; 218:363-7.

137. Santarelli L, Gobbi G, Debs PC et al. Genetic and pharmacological disruption of neurokinin 1 receptor function decreases anxiety-related behaviors and increases serotonergic function. Proc Natl Acad Sci U S A. 2001; 98:1912-7.

138. Kurling S, Kankaanpaa A, Ellermaa S, Karila T, Seppala T. The effect of sub-chronic nandrolone decanoate treatment on dopaminergic and serotonergic neuronal systems in the brains of rats. Brain Res. 2005; 1044:67-75.

139. Kindlundh AM, Bergstrom M, Monazzam A et al. Dopaminergic effects after chronic treatment with nandrolone visualized in rat brain by positron emission tomography. Prog Neuropsychopharmacol Biol Psychiatry. 2002; 26:1303-8.

140. Kindlundh AM, Rahman S, Lindblom J, Nyberg F. Increased dopamine transporter density in the male rat brain following chronic nandrolone decanoate administration. Neurosci Lett. 2004; 356:131-4.

141. Koob GF. Drugs of abuse: anatomy, pharmacology and function of reward pathways. Trends Pharmacol Sci. 1992; 13:177-84.

142. Fuller RW. Fluoxetine effects on serotonin function and aggressive behavior. Ann N Y Acad Sci. 1996; 794:90-7.

143. Muehlenkamp F, Lucion A, Vogel WH. Effects of selective serotonergic agonists on aggressive behavior in rats. Pharmacol Biochem Behav. 1995; 50:671-4.

144. Olivier B, Mos J, van Oorschot R, Hen R. Serotonin receptors and animal models of aggressive behavior. Pharmacopsychiatry. 1995; 2:80-90.

145. Daly RC, Su TP, Schmidt PJ, Pickar D, Murphy DL, Rubinow DR. Cerebrospinal fluid and behavioral changes after methyltestosterone administration: preliminary findings. Arch Gen Psychiatry. 2001; 58:172-7.

146. Strauss RH, Yesalis CE. Anabolic steroids in the athlete. Annu Rev Med. 1991; 42:449-57.

147. Kindlundh AM, Lindblom J, Nyberg F. Chronic administration with nandrolone decanoate induces alterations in the gene-transcript content of dopamine D(1)- and D(2)-receptors in the rat brain. Brain Res. 2003; 979:37-42.

148. Le Greves P, Huang W, Johansson P, Thornwall M, Zhou Q, Nyberg F. Effects of an anabolic-androgenic steroid on the regulation of the NMDA receptor NR1, NR2A and NR2B subunit mRNAs in brain regions of the male rat. Neurosci Lett. 1997; 226:61-4.

149. Le Greves P, Zhou Q, Huang W, Nyberg F. Effect of combined treatment with nandrolone and cocaine on the NMDA receptor gene expression in the rat nucleus accumbens and periaqueductal gray. Acta Psychiatr Scand Suppl. 2002; 412:129-32.

doi:10.1300/J069v25S01_04

Performance Enhancement
and Adverse Consequences of MDMA

Firas H. Kobeissy, MS
Meghan B. O'Donoghue, BS
Erin C. Golden, BS
Stephen F. Larner, PhD
Zhiqun Zhang, MD
Mark S. Gold, MD

SUMMARY. Since its emergence as a recreational club drug in the 1970s, 3-4 methylenedioxy-methamphetamine's usage, more commonly known as ecstasy or MDMA, has risen rapidly through the 1990s, mainly among the teens and young adults of the United States and Europe due to its various perceived social effects. As a result of this sudden rise and the unknown nature of the drug's short- and long-term effects on brain chemistry and function, an ever increasing attention in the political, medical, and scientific fields has occurred which lead to increased mainstream media coverage. Although MDMA is not a performance-enhancing drug in the current use, its abuse is highly associated with motives of euphoria, increased sociability and enhanced energy that can lead to altered cognitive performance. Thus, due to the short-term, acute stimulant effects of MDMA, it is currently viewed by abusers as a performance-enhancing drug affecting both the mood and physical abilities. Given the heightened interest in MDMA as a party drug and in order to shed light on what is currently known about it that leads to its perceived perception as performance enhancing drug, this article will provide a comprehensive overview on MDMA's role as a social facilitator along with its usage history, epidemiology, and its adverse neurotoxic effects. doi:10.1300/J069v25S01_05 *[Article copies available for a fee from The Haworth Document Delivery Service: 1-800-HAWORTH. E-mail address: <docdelivery@haworthpress.com> Website: <http://www. HaworthPress.com> © 2007 by The Haworth Press, Inc. All rights reserved.]*

KEYWORDS. MDMA, ecstasy, club drugs, neurotoxicity, performance enhancing

---

Firas H. Kobeissy is affiliated with the Division of Psychiatry, Meghan B. O'Donoghue, Erin C. Golden, Stephen F. Larner, and Zhiqun Zhang are affiliated with the Department of Neuroscience, and Mark S. Gold is affiliated with the Departments of Neuroscience, Anesthesiology, Community Health and Family Medicine, Division of Addiction Medicine, and Psychiatry; all with McKnight Brain Institute of the University of Florida, Gainesville, FL 32610 USA.

The authors wish to acknowledge the invaluable assistance provided by Ana Amaya, BS, research assistant in Dr. Mark S. Gold Lab, University of Florida, Department of Psychiatry, for the manuscript editing and reviewing without whose help this manuscript could not have been prepared.

[Haworth co-indexing entry note]: "Performance Enhancement and Adverse Consequences of MDMA." Kobeissy, Firas H. et al. Co-published simultaneously in *Journal of Addictive Diseases* (The Haworth Medical Press, an imprint of The Haworth Press, Inc.) Vol. 25, Supplement No. 1, 2007, pp. 47-59; and: *Performance-Enhancing Medications and Drugs of Abuse* (ed: Mark S. Gold) The Haworth Medical Press, an imprint of The Haworth Press, Inc., 2007, pp. 47-59. Single or multiple copies of this article are available for a fee from The Haworth Document Delivery Service [1-800-HAWORTH, 9:00 a.m. - 5:00 p.m. (EST). E-mail address: docdelivery@haworthpress.com].

INTRODUCTION

Ecstasy or 3-4 methylenedioxy-methamphetamine (MDMA), a commonly used illicit drug has become increasingly popular among young people in both the USA and Europe.[1-4] While MDMA is classified as a 'stimulant' like its chemical relative, amphetamine, its effects are less stimulant-like as it promotes an altered sensory, hallucinatory, and emotional experience. MDMA mainly acts as an entactogen and is often used to induce feelings of euphoria, emotional openness, and sensory hallucinations along with the enhanced desire to socialize.[5-7] Due to the social-facilitator effects of MDMA, this drug is often viewed and used as a performance-enhancing agent that helps facilitate social interaction providing a feeling of enhanced energy and emotional freedom. However, MDMA's abuse is not without cost; the negative sequelae associated with MDMA ranges from physiological to neuropsychological deficits.[8] The consequences of this abuse lead to increased accidents, date rapes and reported residual emotional and mental effects long after the drug should have disappeared from body and brain. In humans, MDMA is suspected to have both long and short term toxic effects.[9] These effects include hyperpyrexia, depression and memory loss, in addition to the less well established links to neuronal injury as will be discussed later.[10-12]

While MDMA alters the levels of many of the neurotransmitters in the brain including norepinephrine, dopamine, and acetylcholine, its main effect is to trigger an acute increase followed by a prolonged decrease in normal serotonin levels.[13-15] This sudden shock to the serotonergic system is thought to be responsible for most of the neurotoxic as well as psychological effects of the drug leading to altered brain chemistry and functions.[16,17]

HISTORY, CONSUMPTION, AND EPIDEMIOLOGY

MDMA was first synthesized in 1912 by the German company Merck pharmaceuticals.[18,19] The drug was patented as a tranquillizer in 1960 and as an appetite inhibitor in 1961, but it was not commercially produced and marketed for any of these use.[20,21] In the early 1970s, psychiatrists administered MDMA to their patients to "facilitate psychotherapy" sessions when they noted the drug increased their patients' emotional warmth and reduced their anxiety.[22] However at that time, the Food and Drug Administration had neither approved MDMA for human use nor had any clinical trials been conducted to evaluate its safety or efficacy. Since MDMA offered users pleasurable sensory and emotional experiences at an affordable price, it started appearing on the streets as a recreational drug at about this time.[23]

MDMA is a chemically-synthesized amphetamine which chemically resembles the hallucinogenic substance mescaline where a methylenedioxy $(O-CH_2-O)$ group is attached to the ring structure of its amphetamine molecule.[21] Thus, MDMA pharmacological effects combine that of the amphetamine and mescaline.[21] MDMA is traditionally consumed in tablet forms that are sold in a wide range of colors, shapes, and sizes imprinted with various designs or logos. On the street, MDMA is known as Ecstasy, X, E, Rolls, the Club Drug, Hug Drug, Adam, or Lover's Speed.[7,19,24] Between 1 and 2 tablets are typically taken at rave settings; however, reports indicate that there are individuals who have consumed 10 or more tablets resulting in significant toxic results.[21,25] Like most illicitly prepared drugs, the purity of MDMA varies, but generally, one tablet contains between 50 and 150 mg of MDMA.[21,26] Laboratory analyses of tablets bought on the street show that most of the pills sold as Ecstasy are often not even predominantly made up of MDMA but are cut with other ingredients.[27-29] These may include compounds such as the toxic para-methoxyamphetamine (PMA), 3,4-methylenedioxyamphetamine (MDA), N-ethyl-3-4-methylene dioxyamphetamine (MDEA) and 3,4-methylenedioxy-phenyl-N-methylbutanamine (MBDB) leading to the increased risk of adverse toxic effects upon abuse.[4,21,27,30]

In the mid-1980s, MDMA usage rose significantly and it became one of the most popular drugs among young people under 30 years of age at all-night-dance-parties known as raves.[4,31-33] This sudden increase in abuse propelled both the medical and political communities to question the safety and therapeutic utility of the drug especially because its use was perceived as being safe.[21,34] For two years, hearings were conducted to determine what position the government should assume regarding MDMA.[35] In 1985, finding no therapeutic value for MDMA, the Drug Enforcement Agency (DEA) classified it as a Schedule I controlled substance. In 1986, an administrative judge representing the DEA suggested MDMA be changed to a Schedule III classification, contending that there was adequate evidence to prove its clinical value and safety under strict medical supervision. However, on the recommendation of the DEA, Congress overruled this recommendation and two years later MDMA was permanently classified as a Schedule I controlled substance.[7,19,23,35-38]

While consumption of most illicit drugs in the United States has remained fairly stable, MDMA usage has significantly increased since its first appearance on the street in the mid-1980s. Recent statistics from the U.S. Drug Enforcement Administration show that more than 6.4 million people age 12 and older reported that they have used MDMA at least once in their lives.[31,37] In addition, according to a recent report from the U.S. National Institute of Drug Abuse, it was found that among all 19- to 30-year olds 15% have tried MDMA and among 12th graders that rate was 8% with the life-time prevalence highest at ages 23-24.[39] A similar trend was found in Europe where the popularity of MDMA increased in the 1990s before leveling off to 4-16% of European 15- to 24-year old males reportedly having tried MDMA, and between 2-5% reporting regular use.[36] Furthermore, the 2001 National Household Survey on Drug Abuse reported that in 1992, less than 80,000 individuals aged 12 or older had used MDMA.[40] That number rose to almost two million by the year 2000. In recent years, this upward trend appears to have ceased and actually reversed.[41] Similarly, data presented by the University of Michigan's Monitoring the Future (MTF) study shows that there was a sharp increase in MDMA use from 1999 to 2001, bringing an annual prevalence of 6.2% in 10th graders and 9.2% among 12th graders. Usage also rose among 8th graders in 2000 and 2001, with a decrease in 2002 of about 20% in all three grades and an even a sharper drop in 2003. Such decline continued in all three grades in 2004.[42]

Although still only representing 1% of the total drug-related hospital treatments, the increase in MDMA usage over the past decade is reflected in the increased number of emergency room visits associated with MDMA.[41] From 1999 to 2001 the number of emergency room cases where MDMA was a contributing factor increased 94% as reported by the Drug Abuse Warning Network (DAWN).[43,44] In 2002, the DAWN network study also reported that the collective usage of "club drugs" (GHB, ketamine, LSD, and MDMA) resulted in about 8,100 emergency department visits in 2002, with MDMA accounting for 75% of those visits.[44] However, as is typical with MDMA consumption, 86% of these cases also involved co-intoxication with other drugs. The vast majority of these patients were aged 25 or younger, suggesting that MDMA is primarily affecting the young adult population.[23,31] Finally, a number of fatalities have been associated with MDM usage; seven deaths of young people were registered in 2000 in the state of Florida reportedly caused by the ingestion of MDMA.[27] Interestingly, Patel et al. showed that during the years 1999-2001 a total of 102 deaths were associated with MDMA use across the USA.[4,34]

PHARMACOKINETICS AND METABOLISM

A detailed description of MDMA's pharmacology and pharmacokinetics and metabolism has been extensively studied in a number of reviews.[7,20,21,35,45-47] MDMA metabolism includes its demethylation into either 3,4-methylenedioxyamphetamne (N-demethylation) or into 3,4-dihydroxymethamphetamine (O-demethylation) which is regulated by the liver enzymes, cytochrome P450 3A4 and cytochrome P450 2D6, respectively.[48,49] The enzyme cytochrome P450 2D6 is of unique significance here because approximately 10% of Caucasians have dysfunctional isoforms due to a genetic polymorphism which can reflect on the variable metabolism of MDMA in the population.[9,47,50,51] After ingestion, it was found that MDMA is readily absorbed from the intestinal tract into the bloodstream. It reaches maximum plasma concentrations (T_{max}) two to four hours after oral administration.[52-54] Interestingly, maximum plasma concentrations appear to coincide with the duration of the subjective experience.[9] MDMA passes readily into tissues and much of it is bound to tissue constituents, especially in the brain. Dosage of 50 and 125 mg produces a peak plasma concentration of 106 and 236 ng/ml, respectively.[21] These low concentrations reflect the ability of the drug to enter and bind to various tissues leading to its adverse effects.

As to the pharmacokinetics of MDMA, it has been shown that MDMA has nonlinear pharmacokinetics reflected by a disproportional increase of plasma concentration relative to the dose increase.[52] Thus a small change in dosage may carry some serious risks due to increased toxicity.[47]

It is postulated that this pharmacokinetic profile of increasing plasma concentrations of MDMA, relative to dose ingested, may be due to saturation of metabolic enzymes. Such behavior is attributed to a number of factors including genetic variation of cytochrome p450 enzyme in the population, saturation of different metabolic enzymes and finally due to the various interaction of different MDMA metabolites with different players in metabolic pathways.[9,50] "Street" Ecstasy, consisting of MDMA and its related compounds, is generally produced as a racemic mixture, but it is important to note that these stereoisomers produce different subjective experiences and are metabolized at different rates.[21] S(+) MDMA is preferentially metabolized and cleared, and is more potent than its stereoisomer R(−) MDMA. The S(+) isomer appears to have more amphetamine-like subjective effects, while the R(−) isomer induces subjective effects similar to that of mescaline and LSD.[55,56] As to drug elimination, MDMA has a slow elimination rate with a half life of about 8 hrs and it takes about 40 hrs for 95% clearance from the body.[57] This in part may explain the long term effects of MDMA after 2 days of MDMA ingestion. However some MDMA metabolites, like MDA, have slower elimination rates which relate to the extended MDMA effect long after ingestion.[21]

MDMA USAGE AS A SOCIAL FACILITATOR

It is well established that MDMA is predominantly used for its recreational social facilitating effects that can increase psychomotor drive. The two main reinforcing subjective effects of MDMA abuse include the intensified feeling of attachment and the increased feeling of energy,[9] Furthermore, MDMA is known for its euphoric effects that can keep younger people "in the right mood," thus leading to an altered state of consciousness.[7,58-60] These various effects tend to be mediated by the serotonergic system.[31] These subjective reinforcing effects are summarized in Table 1.

MDMA is a party drug rather than taken by solitary users.[61] Its recreational use is most prevalent among teens and partygoers especially during rave dance parties. Acute subjective reinforcing effects of MDMA include the "Empathy" effect where openness and sharing of emotions are sensed.[62] This is coupled with intense feelings of attachment and connection with heightened sensory perception and increased desire to have sex, although this desire is often not fulfilled since MDMA can inhibit the ability of becoming aroused.[9,19,63] According to a study by Hutton[64] on female clubbers, MDMA was associated with increased feelings of affection; however not necessarily with an increase in sexual drive. Some users actually have a decline in sexual performance.[64,65] In one study conducted in The Netherlands, motives for MDMA use were questioned and evaluated revealing that partygoers are mainly motivated by the energetic and euphoric desirable behavioral effects they perceive or expect upon MDMA use. These included elevated sexiness, self insight, increased energy, extroversion, en-

TABEL 1. Subjective Reinforcing Effects of MDMA

Euphoria
Empathy
Emotional openness
Psychomotor drive
Increased energy
Increased self-confidence
Enhanced mood
Extroversion
Increased sexual drive
Intense feeling of intimacy and connection
Increased sexiness
Heightened self-insight
Intensified feelings
Altered state of mind

hanced mood, and sociability.[6,31,66] In another study evaluating MDMA use by 30 US college students, similar motivational factors related to MDMA use were observed as the ones in the previous study which includes seeking positive effects on mood, desire for an altered state of mind and lessening social pressure. Interestingly, the majority of students were unaware of the adverse effects MDMA could cause.[67] However, the price of such desirable behavioral effects are at the expense of various cognitive tasks deficits including impaired short-term memory, complex attention, difficulty in decision making along with cognition and verbal reasoning as will be discussed later.[68-70]

The enhancing effects that MDMA patients experience are, not surprisingly, similar to those of amphetamines and mescaline. When the non-medical production of MDMA began, it was developed to have a comparable chemical composition to then existing drugs. The effects these drugs cause have also been compared with the effects caused by the natural neurotransmitters epinephrine, dopamine and serotonin.[21] Nevertheless as will be discussed later, unlike methamphetamine, MDMA is thought to not cause dependence.[21]

In the days immediately after MDMA consumption, the mood of the individual declines to below normal, a phenomenon called the "midweek blues."[31] In some instances, this mood reduction results in depression, which in some cases is severe enough to induce suicide.[71] Several studies suggest that this time period is also associated with increased aggression and anger.[31]

MDMA MEDICAL CONSEQUENCES

Characterizing the severity of MDMA abuse in the U.S. is problematic due to the issue of polydrug usage, tablet purity and also due to a scarcity of MDMA clinical data.[4,28,29,41,72] The medical consequences of MDMA are summarized in Table 2. According to Verheyden et al., during the acute phase of intoxication (prior to come-down period), the user experiences unwanted effects such as anxiety, skin irritation along with nausea, arrhythmias, blurred vision, visual hallucinations, motor tics, paresthesias, fainting and decrease respiratory rates, vomit-

TABLE 2. General Medical Consequences of MDMA

Nausea
Vomiting
Headache
Trismus
Bruxism
Hypertension
Palpitations
Anorexia
Urinary incontinence
Difficulty walking
Muscle tension
Blurred vision
Visual hallucinations
Motor tics
Paresthesias
Fainting
Decreased respiratory rate
Hyperthermia/Hypothermia
Pneumomediastinum
Seizures
Status epilepticus
Subarachnoid hemorrhage
Cerebral venous sinus thrombosis

ing, headache, trismus, bruxism, hypertension, palpitations, urinary urgency, muscle aches, and muscle tension which are exacerbated by environmental condition.[11,12,18,59,62,65,73-75] Additionally, the substantial amounts of norepinephrine released by MDMA may cause increased heart rate and blood pressure which can lead to various medical complications.[23] Possible fatal complications as pneumomediastinum, severe chest pain, seizures, status epilepticus, cardiac arrhythmias, asystole, subarachnoid hemorrhage, cerebral venous sinus thrombosis, hyperpyrexia and multi-organ system failure have also been reported. Acute liver damage, caused by the consumption of MDMA, is an important area of concern, with the mechanism of liver damage unclear; although it is believed to be caused by contaminants or metabolites in the preparation.[50,76,77]

Other potentially fatal complications associated with human MDMA use are hyperthermia and dehydration.[7,78-81] Although it has often been assumed that the behaviors associated with MDMA usage (i.e., all night dancing in hot clubs and rave settings) were the main culprit leading to hyperthermia (where body temperature reaches 110°F).[7,81-83] Recent studies indicate that even in the absence of increased motor activity some individuals are still susceptible to hyperthermia.[7,16,78,84] The factors that lead to this hyperthermia are not well understood but they may be related to increased metabolic rate compounded by variations in dose levels or co-intoxication with other drugs, as well as genetic predisposition in relation to metabolism and hormonal secretion.[78,85] It has been shown that genetic predisposition is a major factor mediating MDMA-hyperthermic event. In certain populations, MDMA can stimulate the release of an anti-diuretic hormone–arginine vasopressin (avp)–that retains body water and may lead to taking excess water without sweating and thus can cause hyperthermia and hyponatraemia.[80,86,87]

As to the long-term chronic effects, MDMA has been shown to cause different psychological and psychiatric effects as summarized in Table 3. Due to the serotonergic effect in regulating mood in humans, a number of interesting studies have found some relationship between MDMA use and psychological changes including behavioral impulsivity, memory, mood and cognitive impairment in humans.[10,68,88-91] Psychiatric complications including psychosis, phobic anxiety, panic disorders, and depression have also been studied in MDMA use.[88,89,92,93] More severe levels of paranoia, psychoticism, somatization, anxiety, and hostility are also found in heavy users of ecstasy.[35]

TABLE 3. Psychological and Psychiatric Effects of MDMA

Behavioral impulsivity
Memory impairment
Panic disorders
Depression
Paranoia
Psychoticism
Somatization
Anxiety
Hostility
Problems in learning
Decline in intellectual ability
Decline in executive functioning
Deficits in emotional skills
Impaired set shifting
Deficits in social skills

Considerable declines in general intellectual functioning, memory, learning and executive functioning have been found in MDMA use, with the possibility of full recovery of premorbid levels still unclear as shown by a study conducted by Herkov et al. evaluating MDMA-related neuropsychological functions in MDMA users.[94] These psychological cognitive effects are affected by several factors including but not limited to poly-drug usage.[95] A recent study published by Reay et al.[91] suggests that in poly-drug MDMA usage, individuals who take MDMA in combination with other drugs such as cannabis, cocaine or other stimulants, show a decrease in social and emotional processes. These include process-specific central executive deficits, impaired memory updating and set shifting, as well as deficits in social skills and emotional skills as measured by the Tromo Social Intelligence scale and the emotional questionnaire. All of these processes are located in the orbitoventral region of the prefrontal cortex.[91] However, it is very important to analyze the behavioral/neuropsychological data critically. It is not entirely clear if different psychological changes are purely due to MDMA neurotoxicity, as some Ecstasy users have pre-existing psychopathological conditions as described by Guillot and Greenway.[96] One recent study of recreational Ecstasy use demonstrated that most MDMA users who had been diagnosed with a psychiatric disorder reported being diagnosed prior to Ecstasy use.[96]

There still remains a concern about whether MDMA is addictive. MDMA addictive potential is very low as defined by the Diagnostic and Statistical Manual of Mental Disorders.[21,97] Dependence, however, poses little concern due to the decrease in rewarding effects after repeated use, and an increase in negative side effects with increased usage of the drug. These phenomena provide little incentive for usage frequent enough to be considered a state of dependence. This pattern is similar to a number of other abused hallucinogens, although methamphetamine, another psycho-stimulant, does induce dependence.[21] Rare instances of extreme MDMA abuse, which could be called addiction, have been reported. Jansen described three such cases in 1999.[97] Most of the MDMA users seen by addiction services revealed multiple drug abuse cases which complicate analyses of the addictive role of MDMA.[97,98]

NEUROTOXICITY ASSOCIATED WITH MDMA

It has been shown in animal studies that MDMA exerts neurotoxic effects manifested in the damage of serotonin (5-hydroxytryptamine, 5-HT) axons and nerve terminals in rodents and nonhuman primates in different brain regions; however there is great concern about its possible neurotoxicity in humans.[5,31,99-102] Due to its lipophilic nature, MDMA can readily gain access to the brain. This nature facilitates its diffusion into neuronal cytoplasm and mitochondria enabling it to exert various neurotoxic outcomes. MDMA mainly affects the neurotransmitter systems in the central nervous system. Upon uptake into the brain, MDMA enters the intra-neural space and binds to various types of pre-synaptic transporters affecting the release of a variety of neurotransmitters including serotonin, norepinephrine, and dopamine into the extracellular space. This release stimulates the psychopharmacological effects of MDMA including the temporary euphoria and emotional openness.[7,31,103,104]

Of the three transporters found in neurons, MDMA exhibits the highest affinity for the serotonin transporters (SERTs) where it acts as a substrate.[16,105,106] Upon binding, MDMA leads to a rapid, acute, and reversible elevation in the extracellular concentration of serotonin (5-HT) from the pre-synaptic vesicles. At the same time MDMA blocks the reuptake of 5-HT. Serotonin increase is exacerbated by MDMA's ability to inhibit the two key enzymes responsible for 5-HT metabolism: tryptophan hydroxylase (TPH) and monoamine oxidase (MAO).[7,46,107,108] TPH is the rate-limiting enzyme responsible for serotonin biosynthesis and its activity appears to decline in several parts of the brain including the frontal cortex and hypothalamus. This inhibition allows for an excess synthesis of 5-HT, an effect that can last up to two weeks.[16,109] Similarly, MAO, the main catabolic enzyme responsible for breaking down 5-HT and dopamine is also affected. Its inhibition allows active 5-HT and dopamine to remain elevated within the nerve synaptic

terminal, suggesting that these neurotransmitters remain available for continual release into the extracellular space.[31,107,110]

The exact mechanism by which MDMA mediates its potent effects has yet to be fully identified, however, some of the pathways involved have been elucidated.[46,79,82,104] The level of MDMA neurotoxicity has been found to be dependent on a number of different factors including route of administration, treatment regimens (acute vs. chronic), age, gender, and species of animal studied.[46,111-113] Different studies suggest that MDMA neurotoxicity is attributed to the toxic nature of MDMA's metabolic byproducts including 3,4-dihydroxyamphetamine (HHMA). HHMA can be further metabolized into quinine-like structures, including ortho-quinones, quinine ethers, and N-methyl-alpha-methyl dopamine (MEDA).[114,115] These metabolites are then further oxidized into quinine.[116] The numerous oxidative reactions involved in the production of the final products generate superoxides, hydroxyl radicals, and reactive oxygen species (ROS). Such chemical products are implicated in lipid peroxidation and protein damage which can eventually lead to serotonergic nerve terminal damage.[6,11,16,109] The free radical scavenging drugs, such as sodium ascorbate, were found to be protective against MDMA induced damage, thus supporting the link between ROS formation and MDMA neurotoxicity.[82,109,117] Other studies discuss additional contributing factors including the increase in other neurotransmitter concentrations such as acetylcholine as well as in different hormones including corticosterone and cortisol.[107,118]

Other studies suggest that there is also a significant role for increased extracellular 5-HT production of free radicals which can lead to neurotoxicity.[16,119] The elevation of serotonin involves the stimulation of various receptors and induces activation of different signaling cascades including cAMP and Ca^{2+} resulting in elevated transcription of various genes.[109] The massive serotonin release, along with the reversal of monoamine transporters have been shown to occur in both the cortical and striatal regions. Different MDMA behavioral effects, including hyperthermia and hyperkinesia are also associated with the massive serotonin release and have been shown to be dose-dependent. The effects have been found to occur mainly in the nucleus accumbens and hippocampus.[11,46]

Different animal studies have indicated that MDMA treatment targets different brain regions including the neocortex, hippocampus, and striatum exhibiting alteration in a number of monoaminergic markers.[69] Over the longer term, effects have been shown to occur within 1 day up to one week post-MDMA administration as manifested by a decrease in 5-HT concentrations and a decrease in TPH activity.[31,69]

Along the same line, a number of in vitro studies have been conducted to study the neurotoxic effects of MDMA in tissue and cell culture.[79,104,120-122] In the study conducted by Jimenez et al.,[104] cerebellar granule neuron primary cell cultures (CGNs) treated with 1 to 4 mM MDMA at 24 and 48 hrs time intervals found both an increase in ROS as well as the activation of the caspase-3 cascade. Microscopic evaluation 48 hrs post-MDMA treatment showed neuronal cell damage accompanied by marked increase in ROS production (128%) starting 24 hrs post MDMA treatment. Furthermore, there was 45.3% up-regulation of caspase activity accompanied by nuclear translocation of caspase-3, 48 hours post-MDMA treatment. In addition, caspase activation was also assessed by evaluating α-II spectrin cleavage profile. Interestingly, a marked elevation the 120 kDa α-II spectrin breakdown product was observed. This suggested a prominent role for caspase activation in the apoptotic cascade, an activation that can lead to structural protein proteolysis and neuronal cell death upon MDMA treatment in cell culture.[104,123]

RESEARCH STUDIES

The ability to develop appropriate research models to test the effect of MDMA on humans is hindered by a number of factors. First, it is difficult to develop a model that mimics a regular MDMA user conditions including the dosage, pill purity, multi-drug use and route of administration that typically accompanies consumption. Different models have been developed to study different MDMA effects. The first includes models that seek to evaluate

mechanism(s) of MDMA toxicity. These models study short and long effects on the serotonergic system, including determining the specific MDMA metabolites that cause behavioral, psychological and neurotoxic effects.[124] Other models attempt to replicate the exact typical MDMA user's conditions which include various indices of MDMA usage such as the street drug's impurity, dosage levels, poly-drug usage, age of the user, and cultural conditions of use.[29,125]

The model that studies the negative effects of MDMA on the brain serotonergic system has been well documented by a variety of techniques including in vivo microdialysis, blood plasma analysis, PET imaging and brain tissue assays. These experiments have been performed on numerous species including rats, and nonhuman primates.[6,16,94] In these studies acute biochemical effects appeared as early as 24 hours post consumption and continued for up to one week after MDMA drug administration. In different species, researchers have seen large dose-dependent, reversible depletions of 5-HT and 5-HIAA, as well as a decrease in the activity of TPH with increases in ROS production.[6,16,107] For example, in rats and squirrel monkeys that were given a multi-dose administration of MDMA, there was an initial increase in 5-HT levels but this was followed by an 80% decrease in all parts of the brain within 24 hours. It also became apparent that the acute biochemical effects, particularly the sensitivity to 5-HT depletion was more adverse in nonhuman primates.[16,126]

CONCLUSIONS

Clearly club drugs, including MDMA, are a considerable problem among young people causing a multitude of acute and chronic adverse effects. Their popularity among this demographic group can mostly be attributed to its performance enhancement providing among other things, increased energy and intensified feelings of openness. These drugs can lead to an increase in traffic accidents and the increased chance of medical emergency or death as well as other morbidity. MDMA, as discussed, heightens the likelihood of neurotoxicity which may be overt or may be quite subtle. Brain research findings suggest that when neuronal cells are lost, the young person may appear normal. However, age-related memory decline and cognitive impairment may be more severe in people with pre-existing club drug use as a result of hippocampal and other brain regions cellular loss. Similarly, abusers who perceive MDMA as a performance-enhancing drug to combat social anxiety and enhance social interaction may have cognitive and behavioral effects years later when they are under stress or aging which become manifest because of the loss of cell reserves. Thus, the search for euphoria and heightened energy comes with negative side effects including neuronal loss and even death as discussed in this article. While knowledge about MDMA consumption, metabolism, and appropriate research models has improved, many questions remain to be studied. These include determining which specific metabolites of MDMA can cause cytotoxic effects. How much of neural function has been lost and how much of a decrement from baseline has taken place? Is the MDMA user's demography changing and will this lead to new health consequences for yet unrecognized populations? New research models are still needed to address specific concerns, including the effects on the fetus of pregnant women who use MDMA. More research on these and other questions of MDMA consumption will lead to increased awareness of the consequences of long-term MDMA usage.

In conclusion, this article discusses the reinforcing subjective effects of MDMA perceived as a performance-enhancing drug by the general abusers along with the deleterious adverse effects. The previously described adverse effects of MDMA should prompt us for systematic effective measures at different sectors including researchers, medical personnel and law enforcement groups. In this regard, we should plan for a better monitoring system on epidemiology of the abuse and death associated with MDMA use. Future research including genomics and proteomics should be implemented to describe better the clinical manifestation and mechanisms involved in MDMA abuse. Finally, the vital role of health education and other professional disciplines should be highly stressed to raise awareness and prevention against the risks of various abused drugs.

REFERENCES

1. Baumgarten HG, Lachenmayer L. Serotonin neurotoxins–past and present. Neurotox Res 2004; 6: 589-614.

2. Peroutka SJ. Incidence of recreational use of 3,4-methylenedimethoxymethamphetamine (MDMA, "ecstasy") on an undergraduate campus. N Engl J Med 1987; 317:1542-3.

3. Iannone, M, Bulotta S, Paolino D, Zito MC, Gratteri S, Costanzo FS, Rotiroti D. Electrocortical effects of MDMA are potentiated by acoustic stimulation in rats. BMC Neurosci 2006; 7:13.

4. Schifano F, Corkery J, Deluca P, Oyefeso A, Ghodse AH. Ecstasy (MDMA, MDA, MDEA, MBDB) consumption, seizures, related offences, prices, dosage levels and deaths in the UK (1994-2003). J Psychopharmacol 2006; 20:456-63.

5. Cami J, Farre M, Mas M, Roset PN, Poudevida S, Mas A, San L, de la Torre R. Human pharmacology of 3,4-methylenedioxymethamphetamine ("ecstasy"): psychomotor performance and subjective effects. J Clin Psychopharmacol 2000; 20:455-66.

6. de la Torre R, Farre M, Roset PN, Pizarro N, Abanades S, Segura M, Segura J, Cami J. Human pharmacology of MDMA: pharmacokinetics, metabolism, and disposition. Ther Drug Monit 2004; 26:137-44.

7. Britt GC, McCance-Katz EF. A brief overview of the clinical pharmacology of "club drugs." Subst Use Misuse 2005; 40:1189-201.

8. Rodgers J, Buchanan T, Pearson C, Parrott AC, Ling J, Hefferman TM, Scholey AB. Differential experiences of the psychobiological sequelae of ecstasy use: quantitative and qualitative data from an internet study. J Psychopharmacol 2006; 20:437-46.

9. Gold M, Tabrah H., Frost-Pineda K. Psychopharmacology of MDMA (Ecstasy). Psychiatric Annals 2001; 31:675-81.

10. Able JA, Gudelsky GA, Vorhees CV, Williams MT. 3,4-Methylenedioxymethamphetamine in adult rats produces deficits in path integration and spatial reference memory. Biol Psychiatry 2006; 59(12):1219-26.

11. Cadoni C, Solinas M, Pisanu A, Zernig G, Acquas E, Di Chiara G. Effect of 3,4-methylendioxymethamphetamine (MDMA, "ecstasy") on dopamine transmission in the nucleus accumbens shell and core. Brain Res 2005; 1055:143-8.

12. Mortelmans LJ, Bogaerts PJ, Hellemans S, Volders W, Van Rossom P. Spontaneous pneumomediastinum and myocarditis following Ecstasy use: a case report. Eur J Emerg Med 2005; 12:36-8.

13. Curran HV, Robjant K. Eating attitudes, weight concerns and beliefs about drug effects in women who use ecstasy. J Psychopharmacol 2006; 20:425-31.

14. Schmidt CJ. Neurotoxicity of the psychedelic amphetamine, methylenedioxymethamphetamine. J Pharmacol Exp Ther 1987; 240:1-7.

15. Green AR, Cross AJ, Goodwin GM. Review of the pharmacology and clinical pharmacology of 3,4-

methylenedioxymethamphetamine (MDMA or "Ecstasy"). Psychopharmacology (Berl) 1995; 119:247-60.

16. Green AR, Mechan AO, Elliott JM, O'Shea E, Colado MI. The pharmacology and clinical pharmacology of 3,4-methylenedioxymethamphetamine (MDMA, "ecstasy"). Pharmacol Rev 2003; 55:463-508.

17. Fone KC, Beckett SR, Topham IA, Swettenham J, Ball M, Maddocks L. Long-term changes in social interaction and reward following repeated MDMA administration to adolescent rats without accompanying serotonergic neurotoxicity. Psychopharmacology (Berl) 2002; 159:437-44.

18. Shulgin AT. The background and chemistry of MDMA. J Psychoactive Drugs 1986; 18:291-304.

19. Gold M, Tabrah H. Update on the ecstasy epidemic. Addictions Nursing 2000; 12:133-43.

20. Climko RP, Roehrich H, Sweeney DR, Al-Razi J. Ecstacy: a review of MDMA and MDA. Int J Psychiatry Med 1986; 16:359-72.

21. Kalant H. The pharmacology and toxicology of "ecstasy" (MDMA) and related drugs. Cmaj 2001; 165: 917-28.

22. Greer GR, Tolbert R. A method of conducting therapeutic sessions with MDMA. J Psychoactive Drugs 1998; 30:371-9.

23. Volkow N. MDMA (Ecstasy) Abuse. Research Report Series, Vol. 2006. Rockville, MD: National Institute on Drug Abuse, 2005.

24. NIDA. National Institute on Drug Abuse http://www.drugabuse.gov/clubalert/clubdrugalert.html, 2003.

25. Parrott AC, Lasky J. Ecstasy (MDMA) effects upon mood and cognition: before, during and after a Saturday night dance. Psychopharmacology (Berl) 1998; 139:261-8.

26. Theune M, Esser W, Druschky KF, Interschick E, Patscheke H. [Grand mal series after Ecstasy abuse]. Nervenarzt 1999; 70:1094-7.

27. Goldberger B, Gold M. Ecstasy deaths in the state of Florida: a postmortem analysis. Bio Psychiatry 2002; 51:557.

28. Cole JC, Sumnall HR, Wagstaff GF. What is a dose of ecstasy? J Psychopharmacol 2002; 16:189-90.

29. Bedi G, Redman J. Self-reported ecstasy use: the impact of assessment method on dosage estimates in recreational users. J Psychopharmacol 2006; 20:432-6.

30. Schifano F. New trends in drug addiction: synthetic drugs. Epidemiological, clinical and preventive issues. Epidemiol Psichiatr Soc 2001; 10:63-70.

31. Morton J. Ecstasy: pharmacology and neurotoxicity. Curr Opin Pharmacol 2005; 5:79-86.

32. Schifano F. Potential human neurotoxicity of MDMA ('Ecstasy'): subjective self-reports, evidence from an Italian drug addiction centre and clinical case studies. Neuropsychobiology 2000; 42:25-33.

33. Winstock AR, Griffiths P, Stewart D. Drugs and the dance music scene: a survey of current drug use patterns among a sample of dance music enthusiasts in the UK. Drug Alcohol Depend 2001; 64:9-17.

34. Patel MM, Wright DW, Ratcliff JJ, Miller MA. Shedding new light on the "safe" club drug: methylenedioxymethamphetamine (ecstasy)-related fatalities. Acad Emerg Med 2004; 11:208-10.

35. Valentine G. MDMA and Ecstasy. Psychiatric Times 2002; 19(2).

36. EMCDDA. European Monitoring Centre for Drugs and Drug Addiction The 2005 Annual Report on the state of the drugs problem in Europe European Monitoring Centre for Drugs and Drug Addiction http://www.emcdda.europa.eu/. Luxembourg: European Monitoring Centre for Drugs and Drug Addiction, 2005:44-52.

37. Agency, USDE. MDMA (Ecstasy) Factsheet, Vol. 2006: United States Drug Enforcement Agency.

38. Young FL. In the matter of MDMA scheduling Docket No. 84-48 Opinion and Recommended Ruling, Findings of Fact, Conclusions of Law and Decision of Administrative Law Jude on Issues two through Seven, Vol. 2006: United States Department of Justice, 1986.

39. Johnston L, O'Malley PM., Bachman JG, Schulenbery JE. Monitoring the future national survey results on drug use, 1975-2004. Volume II: College students and adults ages 19-45. Bethesda, MD: National Institue on Drug Abuse, 2005:1-278.

40. NHSDA. National Household Survey on Drug Abuse-SAMHSA Factsheet: http://www.whitehousedrugpolicy.gov/drugfact/nhsda01.html. 2001.

41. Banken JA. Drug abuse trends among youth in the United States. Ann N Y Acad Sci 2004; 1025: 465-71.

42. MTF. Monitoring the Future. National Results on Adolescent Drug use. http://www.monitoringthefuture.org/pubs/monographs/overview2004.pdf: National Institute on Drug Abuse, 2004:1-72.

43. DAWN. Drug Abuse Warning Network (DAWN) http://dawninfo.samhsa.gov/. 2003.

44. Services, USDoHH. Ecstasy, other club drugs & other Hallucinogens, Vol. 2006: SAMSHA, 2006.

45. Rochester JA, Kirchner JT. Ecstasy (3,4-methylenedioxymethamphetamine): history, neurochemistry, and toxicology. J Am Board Fam Pract 1999; 12:137-42.

46. Lyles J, Cadet JL. Methylenedioxymethamphetamine (MDMA, Ecstasy) neurotoxicity: cellular and molecular mechanisms. Brain Res Brain Res Rev 2003; 42:155-68.

47. de la Torre R, Farre M, Ortuno J, Mas M, Brenneisen R, Roset PN, Segura J, Cami J. Non-linear pharmacokinetics of MDMA ('ecstasy') in humans. Br J Clin Pharmacol 2000; 49:104-9.

48. Maurer HH, Bickeboeller-Friedrich J, Kraemer T, Peters FT. Toxicokinetics and analytical toxicology of amphetamine-derived designer drugs ('Ecstasy'). Toxicol Lett 2000; 112-113:133-42.

49. Wu D, Otton SV, Inaba T, Kalow W, Sellers EM. Interactions of amphetamine analogs with human liver CYP2D6. Biochem Pharmacol 1997; 53:1605-12.

50. Tucker GT, Lennard MS, Ellis SW, Woods HF, Cho AK, Lin LY, Hiratsuka A, Schmitz DA, Chu TY. The demethylenation of methylenedioxymethamphet-

amine ("ecstasy") by debrisoquine hydroxylase (CYP2D6). Biochem Pharmacol 1994; 47:1151-6.

51. Gonzalez FJ, Meyer UA. Molecular genetics of the debrisoquin-sparteine polymorphism. Clin Pharmacol Ther 1991; 50:233-8.

52. Mas M, Farre M, de la Torre R, Roset PN, Ortuno J, Segura J, Cami J. Cardiovascular and neuroendocrine effects and pharmacokinetics of 3, 4-methylenedioxy-methamphetamine in humans. J Pharmacol Exp Ther 1999; 290:136-45.

53. Henry JA, Fallon JK, Kicman AT, Hutt AJ, Cowan DA, Forsling M. Low-dose MDMA ("ecstasy") induces vasopressin secretion. Lancet 1998; 351:1784.

54. Verebey K, Alrazi J, Jaffe JH. The complications of 'ecstasy' (MDMA). JAMA 1988; 259:1649-50.

55. Schechter MD. MDMA as a discriminative stimulus: isomeric comparisons. Pharmacol Biochem Behav 1987; 27:41-4.

56. Baker LE, Taylor MM. Assessment of the MDA and MDMA optical isomers in a stimulant-hallucinogen discrimination. Pharmacol Biochem Behav 1997; 57: 737-48.

57. Vereby K, Gold MS. From coca leaves to crack: The effects of dose and routes of adminstration in abuse liability. Psychiatric Annals 1988; 18:513-20.

58. Brookhuis KA, de Waard D, Samyn N. Effects of MDMA (ecstasy), and multiple drugs use on (simulated) driving performance and traffic safety. Psychopharmacology (Berl) 2004; 173:440-5.

59. Sherlock K, Wolff K, Hay AW, Conner M. Analysis of illicit ecstasy tablets: implications for clinical management in the accident and emergency department. J Accid Emerg Med 1999; 16:194-7.

60. Cohen RS. Subjective reports on the effects of the MDMA ('ecstasy') experience in humans. Prog Neuropsychopharmacol Biol Psychiatry 1995; 19: 1137-45.

61. Christopherson A. Amphetamine designer drugs– an overview and epidemiology. Toxicol Lett 2000; 12:127-31.

62. Vollenweider FX, Gamma A, Liechti M, Huber T. Psychological and cardiovascular effects and short-term sequelae of MDMA ("ecstasy") in MDMA-naive healthy volunteers. Neuropsychopharmacology 1998; 19:241-51.

63. Buffum J, Moser C. MDMA and human sexual function. J Psychoactive Drugs 1986; 18:355-9.

64. Hutton F. Up for it, mad for it? Women, drug use and participation in club scenes. Health, Risk and Society 2004; 6: 223-37.

65. Murphy PN, Wareing M, Fisk J. Users' perceptions of the risks and effects of taking ecstasy (MDMA): a questionnaire study. J Psychopharmacol 2006; 20: 447-55.

66. M ter Bogt TF, Engels RC. "Partying" hard: party style, motives for and effects of MDMA use at rave parties. Subst Use Misuse 2005; 40:1479-502.

67. Levy KB, O'Grady KE, Wish ED, Arria AM. An in-depth qualitative examination of the ecstasy experience: results of a focus group with ecstasy-using college students. Subst Use Misuse 2005; 40:1427-41.

68. Quednow BB, Kuhn KU, Hoppe C, Westheide J, Maier W, Daum I, Wagner M. Elevated impulsivity and impaired decision-making cognition in heavy users of MDMA ("Ecstasy"). Psychopharmacology (Berl) 2006: 1-14.

69. Piper BJ, Meyer JS. Memory deficit and reduced anxiety in young adult rats given repeated intermittent MDMA treatment during the periadolescent period. Pharmacol Biochem Behav 2004; 79:723-31.

70. McCardle K, Luebbers S, Carter JD, Croft RJ, Stough C. Chronic MDMA (ecstasy) use, cognition and mood. Psychopharmacology (Berl) 2004; 173:434-9.

71. Iwersen S, Schmoldt A. Two very different fatal cases associated with the use of methylenedioxyethyl-amphetamine (MDEA): Eve as deadly as Adam. J Toxicol Clin Toxicol 1996; 34:241-4.

72. Grob CS, Bravo GL, Walsh RN, Liester MB. The MDMA-neurotoxicity controversy: implications for clinical research with novel psychoactive drugs. J Nerv Ment Dis 1992; 180:355-6.

73. Verheyden SL, Henry JA, Curran HV. Acute, sub-acute and long-term subjective consequences of 'ecstasy' (MDMA) consumption in 430 regular users. Hum Psychopharmacol 2003; 18:507-17.

74. Verheyden SL, Hadfield J, Calin T, Curran HV. Sub-acute effects of MDMA (+/−3,4-methylenedioxy-methamphetamine, "ecstasy") on mood: evidence of gender differences. Psychopharmacology (Berl) 2002; 161:23-31.

75. Siegel RK. MDMA. Nonmedical use and intoxication. J Psychoactive Drugs 1986; 18:349-54.

76. Henry JA, Jeffreys KJ, Dawling S. Toxicity and deaths from 3,4-methylenedioxymethamphetamine ("ecstasy"). Lancet 1992; 340:384-7.

77. Dykhuizen RS, Brunt PW, Atkinson P, Simpson JG, Smith CC. Ecstasy induced hepatitis mimicking viral hepatitis. Gut 1995; 36:939-41.

78. Patel, MM, Belson MG, Longwater AB, Olson KR, Miller MA. Methylenedioxymethamphetamine (ecstasy)-related hyperthermia. J Emerg Med 2005; 29: 451-4.

79. Capela JP, Meisel A, Abreu AR, Branco PS, Ferreira LM, Lobo AM, Remiao F, Bastos ML, Carvalho F. Neurotoxicity of Ecstasy metabolites in rat cortical neurons, and influence of hyperthermia. J Pharmacol Exp Ther 2006; 316:53-61.

80. Nutt DJ. A tale of two Es. J Psychopharmacol 2006; 20:315-7.

81. Dafters RI. Effect of ambient temperature on hyperthermia and hyperkinesis induced by 3,4-methyl-enedioxymethamphetamine (MDMA or "ecstasy") in rats. Psychopharmacology (Berl) 1994; 114:505-8.

82. Bogen IL, Haug KH, Myhre O, Fonnum F. Short- and long-term effects of MDMA ("ecstasy") on synaptosomal and vesicular uptake of neurotransmitters in vitro and ex vivo. Neurochem Int 2003; 43:393-400.

83. Dafters RI. Hyperthermia following MDMA administration in rats: effects of ambient temperature, water consumption, and chronic dosing. Physiol Behav 1995; 58:877-82.

84. Taffe MA, Lay CC, Von Huben SN, Davis SA, Crean RD, Katner SN. Hyperthermia induced by 3,4-methylenedioxymethamphetamine in unrestrained rhesus monkeys. Drug Alcohol Depend 2006; 82:276-81.

85. Freedman RR, Johanson CE, Tancer ME. Thermoregulatory effects of 3,4-methylenedioxymethamphetamine (MDMA) in humans. Psychopharmacology (Berl) 2005; 183:248-56.

86. Fallon JK, Shah D, Kicman AT, Hutt AJ, Henry JA, Cowan DA, Forsling M. Action of MDMA (ecstasy) and its metabolites on arginine vasopressin release. Ann N Y Acad Sci 2002; 965:399-409.

87. Forsling ML, Fallon JK, Shah D, Tilbrook GS, Cowan DA, Kicman AT, Hutt AJ. The effect of 3,4-methylenedioxymethamphetamine (MDMA, 'ecstasy') and its metabolites on neurohypophysial hormone release from the isolated rat hypothalamus. Br J Pharmacol 2002; 135:649-56.

88. Soar K, Turner JJ, Parrott AC. Psychiatric disorders in Ecstasy (MDMA) users: a literature review focusing on personal predisposition and drug history. Hum Psychopharmacol 2001; 16:641-5.

89. Thomasius R, Petersen KU, Zapletalova P, Wartberg L, Zeichner D, Schmoldt A. Mental disorders in current and former heavy ecstasy (MDMA) users. Addiction 2005; 100:1310-9.

90. Morgan MJ, McFi, L, Fleetwood H, Robinson JA. Ecstasy (MDMA): are the psychological problems associated with its use reversed by prolonged abstinence? Psychopharmacology (Berl) 2002; 159:294-303.

91. Reay JL, Hamilton C, Kennedy DO, Scholey AB. MDMA polydrug users show process-specific central executive impairments coupled with impaired social and emotional judgement processes. J Psychopharmacol 2006; 20:385-8.

92. Parrott, AC. Is ecstasy MDMA? A review of the proportion of ecstasy tablets containing MDMA, their dosage levels, and the changing perceptions of purity. Psychopharmacology (Berl) 2004; 173:234-41.

93. Dughiero G, Schifano F, Forza G. Personality dimensions and psychopathological profiles of Ecstasy users. Hum Psychopharmacol 2001; 16:635-9.

94. Herkov M, Gold M, Frost-Pineda K. Neuropsychological consequences of MDMA. Biol Psychiatry 2001; 49:1S-176S.

95. Parrott AC. Human psychopharmacology of Ecstasy (MDMA): a review of 15 years of empirical research. Hum Psychopharmacol 2001; 16:557-77.

96. Guillot C, Greenway D. Recreational ecstasy use and depression. J Psychopharmacol 2006; 20:411-6.

97. Jansen KL. Ecstasy (MDMA) dependence. Drug Alcohol Depend 1999; 53:121-4.

98. Schifano F, Di Furia L, Forza G, Minicuci N, Bricolo R. MDMA ('ecstasy') consumption in the context of polydrug abuse: a report on 150 patients. Drug Alcohol Depend 1998; 52:85-90.

99. Ricaurte GA, McCann UD. Assessing long-term effects of MDMA (Ecstasy). Lancet 2001; 358:1831-2.

100. Commins, DL, Vosmer G, Virus RM, Woolverton WL, Schuster CR, Seiden LS. Biochemical and histological evidence that methylenedioxymethylamphetamine (MDMA) is toxic to neurons in the rat brain. J Pharmacol Exp Ther 1987; 241:338-45.

101. O'Hearn E, Battaglia G, De Souza EB, Kuhar MJ, Molliver ME. Methylenedioxyamphetamine (MDA) and methylenedioxymethamphetamine (MDMA) cause selective ablation of serotonergic axon terminals in forebrain: immunocytochemical evidence for neurotoxicity. J Neurosci 1988; 8:2788-803.

102. Green AR, Goodwin GM. Ecstasy and neurodegeneration. BMI 1996; 312:1493-4.

103. Salzmann J, Marie-Claire C, Noble F. Acute and long-term effects of ecstasy. Presse Med 2004; 33:24-32.

104. Jimenez A, Jorda EG, Verdaguer E, Pubill D, Sureda FX, Canudas AM, Escubedo E, Camarasa J, Camins A, Pallas M. Neurotoxicity of amphetamine derivatives is mediated by caspase pathway activation in rat cerebellar granule cells. Toxicol Appl Pharmacol 2004; 196:223-34.

105. Steele TD, Nichols DE, Yim GK. Stereochemical effects of 3,4-methylenedioxymethamphetamine (MDMA) and related amphetamine derivatives on inhibition of uptake of [3H]monoamines into synaptosomes from different regions of rat brain. Biochem Pharmacol 1987; 36:2297-303.

106. Battaglia G, Brooks BP, Kulsakdinun C, De Souza EB. Pharmacologic profile of MDMA (3,4-methylenedioxymethamphetamine) at various brain recognition sites. Eur J Pharmacol 1988; 149:159-63.

107. Cole JC, Sumnall HR. The pre-clinical behavioural pharmacology of 3,4-methylenedioxymethamphetamine (MDMA). Neurosci Biobehav Rev 2003; 27: 199-217.

108. Bialer PA. Designer drugs in the general hospital. Psychiatr Clin North Am 2002; 25:231-43.

109. Thiriet N, Ladenheim B, McCoy MT, Cadet JL. Analysis of ecstasy (MDMA)-induced transcriptional responses in the rat cortex. Faseb J 2002; 16:1887-94.

110. Sulzer D, Sonders MS, Poulsen NW, Galli A. Mechanisms of neurotransmitter release by amphetamines: a review. Prog Neurobiol 2005; 75:406-33.

111. De Souza EB, Battaglia G, Insel TR. Neurotoxic effect of MDMA on brain serotonin neurons: evidence from neurochemical and radioligand binding studies. Ann N Y Acad Sci 1990; 600:682-97; discussion 697-8.

112. McKenna DJ, Guan XM, Shulgin AT. 3,4-Methylenedioxyamphetamine (MDA) analogues exhibit differential effects on synaptosomal release of 3H-dopamine and 3H-5-hydroxytryptamine. Pharmacol Biochem Behav 1991; 38:505-12.

113. Broening HW, Bowyer JF, Slikker W, Jr. Age-dependent sensitivity of rats to the long-term effects of

the serotonergic neurotoxicant (±)-3,4-methylenedi-oxymethamphetamine (MDMA) correlates with the magnitude of the MDMA-induced thermal response. J Pharmacol Exp Ther 1995; 275:325-33.

114. Bai F, Jones DC, Lau SS, Monks TJ. Serotonergic neurotoxicity of 3,4-(±)-methylenedioxy-amphetamine and 3,4-(±)-methylendioxymethampheta-mine (ecstasy) is potentiated by inhibition of gamma-glutamyl transpeptidase. Chem Res Toxicol 2001; 14:863-70.

115. Bai F, Lau, SS, Monks TJ. Glutathione and N-acetylcysteine conjugates of alpha-methyldopamine produce serotonergic neurotoxicity: possible role in methylenedioxyamphetamine-mediated neurotoxicity. Chem Res Toxicol 1999; 12:1150-7.

116. Hiramatsu M, Kumagai Y, Unger SE, Cho AK. Metabolism of methylenedioxymethamphetamine: formation of dihydroxymethamphetamine and a quinone identified as its glutathione adduct. J Pharmacol Exp Ther 1990; 254:521-7.

117. Gudelsky GA. Effect of ascorbate and cysteine on the 3,4-methylenedioxymethamphetamine-induced depletion of brain serotonin. J Neural Transm 1996; 103:1397-404.

118. Laviola G, Adriani W, Terranova ML, Gerra G. Psychobiological risk factors for vulnerability to psychostimulants in human adolescents and animal models. Neurosci Biobehav Rev 1999; 23:993-1010.

119. Seiden LS, Sabol KE. Methamphetamine and methylenedioxymethamphetamine neurotoxicity: possible mechanisms of cell destruction. NIDA Res Monogr 1996; 163:251-76.

120. Won L, Bubula N, Heller A. Fetal exposure to (±)-methylenedioxymethamphetamine in utero enhances the development and metabolism of serotonergic neurons in three-dimensional reaggregate tissue culture. Brain Res Dev Brain Res 2002; 137:67-73.

121. Tiangco DA, Lattanzio FA, Jr., Osgood CJ, Beebe SJ, Kerry JA, Hargrave BY. 3,4-Methylenedioxy-methamphetamine activates nuclear factor-kappaB, increases intracellular calcium, and modulates gene transcription in rat heart cells. Cardiovasc Toxicol 2005; 5:301-10.

122. Hrometz SL, Brown AW, Nichols DE, Sprague JE. 3,4-Methylenedioxymethamphetamine (MDMA, ecstasy)-mediated production of hydrogen peroxide in an in vitro model: the role of dopamine, the serotonin-reuptake transporter, and monoamine oxidase-B. Neurosci Lett 2004; 367:56-9.

123. Cadet JL, Jayanthi S, Deng X. Methamphetamine-induced neuronal apoptosis involves the activation of multiple death pathways. Review. Neurotox Res 2005; 8:199-206.

124. Escobedo I, O'Shea E, Orio L, Sanchez V, Segura M, de la Torre R, Farre M, Green AR, Colado MI. A comparative study on the acute and long-term effects of MDMA and 3,4-dihydroxymethamphetamine (HHMA) on brain monoamine levels after i.p. or striatal administration in mice. Br J Pharmacol 2005; 144:231-41.

125. Mechan, A, Yuan, J, Hatzidimitriou G, Irvine RJ, McCann UD, Ricaurte GA. Pharmacokinetic profile of single and repeated oral doses of MDMA in squirrel monkeys: relationship to lasting effects on brain serotonin neurons. Neuropsychopharmacology 2006; 31:339-50.

126. Hatzidimitriou G, McCann UD, Ricaurte GA. Altered serotonin innervation patterns in the forebrain of monkeys treated with (+/−)3,4-methylenedioxy-methamphetamine seven years previously: factors influencing abnormal recovery. J Neurosci 1999; 19:5096-107.

doi:10.1300/J069v25S01_05

Performance Enhancing, Non-Prescription Use of Erectile Dysfunction Medications

Noni A. Graham, MPH
Alexandria (Alexis) Polles, MD
Mark S. Gold, MD

SUMMARY. Erectile dysfunction (ED) has a variety of causes, e.g., traumatic injuries, surgery in the pelvic region, use of certain drugs, prostate cancer, diabetes, hypertension, and kidney diseases. The advent of several oral therapies for treatment of erectile dysfunction have afforded many men a relatively quick and easy alternative to previously high-risk, invasive treatments, e.g., intracavernosal and intraurethral insertions, vascular surgery, vacuum constriction devices, and prostheses. However, ED drugs are being misused by men and women of all ages, with documented use in college-aged individuals and homosexual men, many times without a prescription. Furthermore, we have collected data on sexual compulsivity patients which confirm the abuse potential of ED medications among certain psychiatric patients. Those with depression, cardiovascular disease, and nicotine addiction, are also at increased risk of misuse. Viagra and other ED medications can easily be obtained online, on the street, in sex shops, etc., and are frequently used concomitantly with recreational drugs, e.g., MDMA, GHB, amyl nitrites, methamphetamine, and ketamine. Currently, no national survey or surveillance system tracks misuse of any erectile dysfunction medication. doi:10.1300/J069v25S01_06 *[Article copies available for a fee from The Haworth Document Delivery Service: 1-800-HAWORTH. E-mail address: <docdelivery@haworthpress.com> Website: <http://www. HaworthPress.com> © 2007 by The Haworth Press, Inc. All rights reserved.]*

KEYWORDS. Abuse, Cialis, compulsive, erectile dysfunction, impotence, Levitra, misuse, nitric oxide, non-prescription, performance, sexual, UPRIMA, Viagra, Yohimbe

IMPOTENCE

According to a Consensus statement by the National Institutes of Health, impotence or erectile dysfunction (ED) refers to the inability of a male to achieve an erect penis as part of the multifaceted process of male sexual function, i.e., desire, orgasmic capability, ejaculatory ca-

Noni A. Graham is Coordinator, Department of Psychiatry, Division of Addiction Medicine, University of Florida College of Medicine, P.O. Box 100183, Gainesville, FL 32610.

Alexandria (Alexis) Polles is Medical Director, Pine Grove Behavioral Health and Addiction Services Gentle Path Program, 304 Emerald Lane, Hattiesburg, MS 39404.

Mark S. Gold is Distinguished Professor, Departments of Psychiatry, Neuroscience, Anesthesiology, Community Health and Family Medicine, Chief, Division of Addiction Medicine, University of Florida College of Medicine, P.O. Box 100183, Gainesville, FL 32610.

Address correspondence to: Noni A. Graham at the above address (E-mail: NoniG@psychiatry.ufl.edu).

[Haworth co-indexing entry note]: "Performance Enhancing, Non-Prescription Use of Erectile Dysfunction Medications." Graham, Noni, A., Alexandria (Alexis) Polles, and Mark S. Gold. Co-published simultaneously in *Journal of Addictive Diseases* (The Haworth Medical Press, an imprint of The Haworth Press, Inc.) Vol. 25, Supplement No. 1, 2007, pp. 61-68; and: *Performance-Enhancing Medications and Drugs of Abuse* (ed: Mark S. Gold) The Haworth Medical Press, an imprint of The Haworth Press, Inc., 2007, pp. 61-68. Single or multiple copies of this article are available for a fee from The Haworth Document Delivery Service [1-800-HAWORTH, 9:00 a.m. - 5:00 p.m. (EST). E-mail address: docdelivery@haworthpress.com].

pacity. In the U.S. approximately 30 million men suffer from ED. Previously assumed to be a natural event in the process of aging, we now know that erectile dysfunction has a variety of causes and can affect men of all ages. The condition may stem from illnesses, e.g., diabetes, hypertension, kidney diseases, prostate cancer; injuries, including spinal cord trauma; habits, e.g., alcohol and tobacco use; even medical treatments, polypharmacy, or surgeries in the pelvic region. Many people also fail to realize the role of the brain in sexual satisfaction. Psychological problems like depression, fear and anxiety due to poor self-image, lack of sexual knowledge, or a breakdown of interpersonal relationships frequently lead to the inability to develop or maintain an erection. Aside from psycho-behavioral therapy, which has not been studied for ED in detail, treatment options for ED have been high-risk and invasive. Historically, ED has been treated with hormone replacement therapy, intracavernosal injections, intraurethral insertions, vacuum/constrictive devices, vascular surgery, and penile prostheses.[1] For the above reasons, or being unwilling to comfortably discuss intimate problems with a stranger (among other reasons), many sufferers of ED failed to seek or initiate treatment for this common problem.

THE DEVELOPMENT OF VIAGRA

In the 80s, Ferid Murad was the first to discover that nitroglycerin releases nitric oxide, a colorless gas which causes smooth muscle to relax. Murad, along with Robert Furchgott and Louis Ignarro, expanded upon this newfound knowledge and declared nitric oxide regulates blood vessels and could be used to treat heart disease and shock. By 1991, Andrew Bell, David Brown, and Nicholas Terrett of the Pfizer Laboratories of Kent, England found that the pyrazolopyrimidone class of compounds would help in treating angina. Thus, Sildenafil became patented and underwent clinical trials. In 1992, the trials were ceased after unsuccessful results in heart patients. However, within the trials, Peter Ellis and Nicholas Terrett noticed that Sildenafil increased blood flow to the penis by the enhancement of nitric oxide. After further research, it was decided that Sildenafil be

marketed as a new treatment for erectile dysfunction. Soon after, Peter Dunn and Albert Wood developed a 9-step process to mass-produce sildenafil citrate in pill form under the trade name Viagra.[2]

Prescribing Information

Approved for sale by the FDA only 6-months after submission, on March 27th, 1998, Viagra became the first non-invasive, non-surgical, cost-effective treatment for erectile dysfunction or impotence, enabling men to respond to sexual stimulation. The prescription drug works as a selective inhibitor of cyclic guanosine monophosphate (cGMP)-specific phosphodiesterase type 5 (PDE5), increasing blood flow to the penis so that a man may consistently get and keep an erection when excited. Sildenafil citrate itself is a whitish, crystalline powder, and is prepared as a blue, rounded-diamond-shaped, fast-acting pill, shown to improve erections in approximately 80% of men. Viagra has a median absorption time of 60 minutes and a terminal half life of approximately 4 hours. It can be taken once daily an hour before sexual activity and is available in 3 doses: 25 mg, 50 mg, and 100 mg. Currently it is used by at least 23 million men worldwide.

Normally, during an erection nitric oxide is released into the corpus cavernosum, activating guanylate cyclase to increase levels of cyclic guanosine monophosphate (cGMP). This produces smooth muscle relaxation and inflow of blood. Viagra works to enhance the effect of nitric oxide. Clinical trials revealed decreases in supine blood pressure approximately 1-2 hours after ingestion, thus, Viagra should not be taken with nitrate drugs as it can lead to hypopiesis. Viagra is not recommended for women or anyone under 18 years of age.[3]

Proven Success

In 2001 in a study of over 1000 participants, Steers et al. reported long-term safety and efficacy of flexible dose Viagra treatment with 92% and 89% of patients reporting improved erections at 36 and 52 weeks, respectively.[4]

Paige et al. have reported on the success of Viagra in many facets of life.[5] Using the International Index of Erectile Function and the marital interaction scale from the Cancer Rehabilitation Evaluation System, Sildenafil users were found to have an increase in erectile function, overall sexual satisfaction, and intercourse satisfaction (p < 0.001). Sexual partner relationship and emotional well-being were also improved (p = 0.007 and p < 0.001, respectively). Participants expressed that Viagra improved their quality of life and their relationships with their partners.[5]

According to Muller and Benkert, Sildenafil users also report a decrease in depressive symptoms.[6] In a cross-sectional study of men diagnosed with ED, successful Viagra users were compared to men awaiting ED treatment using the Center of Epidemiologic Studies Depression Scale (CES-D). Apparently, 16% of those awaiting treatment suffered depressive symptoms, compared to 4% of those with successful experiences with sildenafil. The sildenafil treatment group had positive affect and hedonistic drive in comparison to those awaiting treatment.[6]

Novel Uses for Viagra

Interestingly, Nurnberg et al. report on a sample of women aged 18-60 who were experiencing sexual dysfunction for more than 4 weeks due to antidepressants.[7] Upon taking 1-2 50 mg tablets of Sildenafil, 9 of 10 women reported reversal of sexual dysfunction: effective duration and intensity of adequate arousal, lubrication, and orgasmic function.[7]

Viagra may also become choice therapy in adults with pulmonary hypertension. Inhaled nitric oxide (INO) is an expensive, investigative treatment requiring continuous inhalation via a closed breathing circuit, and has been used for a variety of ailments, including acute respiratory distress syndrome, severe hypoxemia, cardiac failure, and pulmonary hypertension. This intricate treatment modality prompted researchers to address delivery of nitric oxide in other forms, including sustained treatment outside of the healthcare facility. In a case report by Ng et al., a 73-yr old morbidly obese woman presented to the ICU with a variety of symptoms, including nephritis, hyperthyroidism, metabolic acidosis, and respiratory arrest.[8] Upon confirmation of pulmonary hypertension and epinephrine and INO treatment, the patient was

weaned with 25 mg of Viagra daily, which immediately decreased her mean arterial pressure and mean pulmonary artery pressure. Her overall condition improved with no adverse effects, allowing her to return home 2 1/2 weeks later.[8]

Sildenafil may be helpful in treatment of Crohn's disease. The Crohn's & Colitis Foundation of America define Crohn's disease as a chronic immune disorder causing inflammation of the gastrointestinal tract. The immune system mistakes the body's natural flora for foreign substances, sending macrophages to the gastrointestinal tract thus, producing inflammation in those areas.[9] Marks et al. believe a weakened immune response (especially of neutrophils) and poor blood flow add to this inflammation.[10] In a small study of patients with Crohn's disease and controls, the research team (Marks et al.) from University College London (UCL) used biopsies from the bowel and skin from the arm to quantify neutrophil recruitment and cytokine production after acute trauma to these areas, also measuring interleukin 8 secretion by cultured monocyte-derived macrophages after exposure to inflammatory mediators. To understand local inflammatory response and vascular changes, the patients were given injections of heat-killed *Escherichia coli*. In the controls, inflammation was associated with production of nitric oxide. Use of Viagra in Crohn's sufferers mediated the poor blood flow, thereby reducing inflammation.[10]

Adverse Effects of Viagra

The main side effects of Viagra include headache, facial flushing, upset stomach, nasal congestion, urinary tract infection, bluish or blurred vision, sensitivity to sunlight, diarrhea, dizziness, rash, or rarely, erections lasting longer than 4 hours. In some persons, Viagra may cause cardiovascular and cerebrovascular events, including angina, myocardial infarction and hemorrhaging. Many suffer seizures and anxiety, anemia, hematuria and epistaxis. Some may suffer various ailments of the eye: diplopia, non-arteritic anterior ischemic optic neuropathy (NAION), blindness, ocular burning and swelling, retinal bleeding, and vitreous detachment.[3]

In the UK, Akash et al. report on a 54-year-old man who developed blindness in his left eye after a Viagra overdose.[11] Apparently he took 200 mg a few hours before the occurrence–his usual dose was 100 mg 2-3 times per week–the maximum allowed. After discovery of a relative afferent pupillary defect, he was diagnosed with left combined NAION with cilioretinal artery occlusion. He has discontinued use of Viagra and has been placed on 75 mg of aspirin a day to prevent fellow eye involvement. Unfortunately there are no treatments for NAION at this time.[11]

ALTERNATIVE THERAPIES FOR ERECTILE DYSFUNCTION

In the last 2 decades, several drugs have been marketed for ED (Table 1).

Intracavernosal injections CAVERJECT Impulse and EDEX (Alprostadil or prostaglandin E_1) were released in 1995 and 1996, respectively. These smooth muscle relaxants allow for the corporal veno-occlusive mechanism to occur–the lacunar spaces expand and blood is entrapped through compression of venules against the tunica albuginea, thus, providing an erection.[12,13]

MUSE (medicated urethral system for erection) is an alprostadil intraurethral insertion. A tube is inserted into the urethra to release the medicated pellet locally with results in about 10 minutes.[14] UPRIMA or amorphine hydrochloride hemyhydrate sublingual tablets are dopamine receptor agonists that work in the hypothalamic region of the brain to aid in mediation of erection.[15]

Levitra or vardenafil HCl tablets were released in 2003 by Bayer Aktiengesellschaft in conjunction with GlaxoSmithKline and Schering-Plough Corporation. A selective inhibitor of cGMP-specific PDE5 (like Viagra), vardenafil HCl should be taken only once per day; one hour before sexual activity. Similar to Viagra, the mean terminal half-life of Levitra and its metabolites is 4-5 hours.[16]

Also released in 2003, Lilly ICOS' Cialis or tadalafil follows the same mechanisms as Viagra and Levitra: selective inhibition of cGMP-specific PDE5. Interestingly, Cialis tablets can work in as little as 30 minutes, and the mean terminal half-life of tadalafil is 17.5 hours; 4 times that of both Viagra and Levitra.[17]

Yohimbe (a prescription medication and herbal supplement) is one of the most popular formulations used for sexual enhancement. Termed a natural aphrodisiac, it has been used to reduce prostate inflammation, treat menstrual pain, and even as a local anesthetic for minor surgery. Extracted from the bark of West African evergreen *Pausinystalia yohimbe*, the yohimbe alkaloid is a potent alpha-2-adrenergic antagonist that increases adrenal output of norepinephrine, thus increasing blood flow to the penis.[18] Yohimbine hydrochloride tablets are marketed under several trade names; Actibine, Baron-X, Prohim, Viritab, Yocon, Yohimex, Yoman, etc., used mostly for its ability to increase blood flow and dilate pupils.[19]

TABLE 1. Comparison of Several Erectile Dysfunction Drugs

NAME	ROUTE	TIME NEEDED FOR RESULTS	FDA APPROVED?	LIKELIHOOD OF MISUSE
Viagra®	oral	1 hour	Yes	high
Cialis®	oral	30 minutes	Yes	high
Levitra®	oral	1 hour	Yes	high
Caverject®	injection	5 minutes	Yes	low
Edex®	injection	5 minutes	Yes	low
Muse®	suppository	10 minutes	Yes	medium
Uprima®	oral	10 minutes	Yes	very high
Yohimbe/Yohimbine HCl	oral	2 weeks	Yes	high

For those who have not been successful with any of the aforementioned pharmaceuticals, there are hosts of "natural" or "herbal" formulations: pills, creams, gels, patches, and injections that claim similar results without as many side effects. However, these formulations are not regulated by the FDA and only tested on animals; thus, these alternatives should be used with extreme care.

NON-PRESCRIPTION USE

Prescription misuse is on the rise. According to the 2005 Monitoring the Future Survey, 12th graders are using more tranquilizers and sedatives, with use increasing to 7% this year from 2.8% in 1992.[20] Currently, no national survey or surveillance system tracks misuse of any erectile dysfunction medication. According to an article in press, sales of ED drugs have been falling. Apparently, doctors wrote 10% more prescriptions in October 2004 than October 2005. Although 50% of men over 40 years of age suffer from some form of ED, only 15% of these men received a prescription for ED drugs like Viagra.[21] How are the rest of these men accessing ED drugs? There are documented, reputable websites which allow for Viagra to be prescribed and ordered (even in bulk) without a face-to-face meeting with a physician. Many people want to retain anonymity and believe ordering through the internet will help them to do so. Although less expensive than surgical interventions or injections, Viagra still costs at least $10-$20 per pill. In a nation where the majority is without health insurance, people search for cheaper alternatives which often may be less efficacious. Many people acquire the drug illegally from a number of sources, including online retailers who may sell cheap imitations, sex shops, dealers, etc. Pfizer mentions 4 common imitations: Counterfeit Viagra (made to look like Viagra but sold cheaply), Generic Viagra (the FDA has not approved a generic form of Viagra), Quick-dissolving Viagra (the FDA has not approved a quick-dissolving form of Viagra), and Herbal medicines (these are not tested clinically in humans and could be contaminated). The following describes instances of nonmedical Viagra use in special populations. Although there are no studies in the literature to date, it would be interesting to understand nonprescription use of erectile dysfunction drugs among men and women under 40, and those who use these drugs to self-medicate, e.g., depressed or alcohol and nicotine dependent individuals.

Homosexual Men

According to Crosby and DiClemente, in a survey of men having sex with men (MSM), 37% of cocaine users and 35% of ecstasy users were also Viagra users.[22] Sixteen percent of the total sample reported use of non-prescription Viagra.[22] In another study by Chu et al., among MSM who averaged 35 years of age, 32% reported ever using Viagra while 21% used it in the previous 6 months.[23] Recent use was found to be associated with unprotected anal sex, being HIV-positive, increasing age, illicit drug use, and white ethnicity. Some probands used Viagra concomitantly with speed (23%), ecstasy (18%), amyl nitrite/poppers (18%), Highly Active Anti Retroviral Treatment (11%), ketamine (11%), and gamma-hyrdroxybutyrate (8%). Forty-four percent of Viagra users obtained it from a friend, 6% from the internet, 4% on the street, and 10% from undisclosed sources.[23]

In a survey of men seeking public STD services in San Francisco, gay or bisexual men were more likely to use Viagra than heterosexual men ($p < 0.01$). Most of these individuals (56%) reported receiving Viagra from a friend, but also reported receiving the drug from a healthcare provider (42%) or the internet (2%). Similar to the aforementioned results, among gay and bisexual men, Viagra was reported to be used in conjunction with ecstasy (43%), methamphetamines (28%), amyl nitrate or poppers (15%), and ketamine (8%). Among gay and bisexual users who mixed Viagra with other drugs, 69% were less than 35 years of age and 73% received Viagra from a friend and believed mixing Viagra with other drugs would enhance their sexual experience. Again, Viagra use in the last year was more common among HIV-positive individuals than those who were HIV-negative ($p = 0.1$). Gonorrhea, Chlamydia, and Syphilis infections were greater among Viagra users than nonusers ($p = 0.09$) and the HIV-positive Viagra users were more likely to have these infections than HIV-positive nonusers (p = 0.05). Viagra users also had a higher number of sex partners and recent unprotected sex with partners of opposite or unknown HIV status (p < 0.01 and p = 0.03, respectively).[24]

Furthermore, upon evaluation of predictors of sexual risk behaviors and diseases among males with HIV infection, Cachay, Mar-Tang, and Matthews found potentially transmitting sexual risk behavior to be associated with use of drugs or alcohol during sex, white race, male partners, and Viagra use.[25] Apparently, drug abuse happens most frequently during circuit parties (CPs), where gay and bisexual men congregate for parties and other social events, typically lasting through a weekend. In these instances, Colfax et al. report that crystal methamphetamine, amyl nitrite, and Viagra users were more likely to engage in risky sexual behaviors, i.e., unprotected anal sex with partners with unknown HIV status.[26] One-fifth of the entire sample was HIV positive, but one-third of Viagra users were HIV positive.[26]

College-Aged Individuals

According to Delate, Simmons, and Motheral, in profiling commercially insured adult beneficiaries, the prevalence of Viagra use increased 84% since 1998.[27] Apparently the fastest growing segment of Viagra users are not older males as would be suspected, but males, 18-45 years of age as well as a growing young female population. The researchers suggest the younger patients are using Viagra without any underlying etiologic reasons.[27]

In a study by Aldridge and Measham, soon after its release in England, Viagra was used recreationally by healthy young people in the night clubs of England.[28] Reportedly the drug was combined with a variety of drugs, including methylenedioxymethamphetamine (MDMA or ecstasy), cocaine, cannabis, amyl nitrate, gamma-hydroxybutyric acid, and alcohol. Viagra gave respondents a feeling of 'warmth,' enhancing sexual desire and love making. They obtained Viagra from friends, dealers, sex shops, and the internet.[28]

Despite eminent gossip, Viagra has not been found to improve sexual function in men without erectile dysfunction. A randomized, double-blind, placebo-controlled trial among men aged 20-40, showed no differences in improvement of erection quality. However, during continued erotic stimulus, some Viagra users were able to obtain subsequent erections in less time post-ejaculation (x^2 test, p = 0.04).[29]

Contrary to popular belief, ED drugs are not aphrodisiacs; they will only work if the user is erotically stimulated; e.g., Viagra, Cialis, and Levitra work in the presence of nitric oxide which is released locally upon arousal.

Compulsive Sex Addicts

There is evidence that patients with problematic, compulsive on-line sexual behavior are concerned about co-occurring sexual performance dysfunction. Our preliminary data from Pine Grove Behavioral Health and Addiction Services reveal prescription use and misuse in a treatment-seeking sample of male patients with sexual compulsivity disorders. Of 41 newly admitted cases, 95% were White and 5% were Hispanic with ages varying significantly: 21-66 years of age (average age = 42). Approximately 48.8% reported current use of some type of prescription medication for erectile dysfunction. Exactly half of these individuals reported having a prescription, although at least one patient admitted to writing prescriptions for himself. Thirty percent reported obtaining the drugs by "other" means, e.g., "off the street." The remaining patients obtained ED drugs from the "internet" (15%) or "from a friend" (5%). Sixty percent mentioned use of only Viagra, while 20% used both Viagra and Cialis, 10% used Viagra and Levitra, and 10% used Viagra, Levitra, and Cialis.[30]

CONCLUSION

Recently developed, less invasive oral therapies like Viagra have made it easier for millions of men to combat erectile dysfunction due to illness, trauma, surgery, medications, and aging. The ability to fill prescriptions for many of these drugs online may be helpful for men who wish to retain anonymity and avoid stigma associated with suffering from ED. However, this method of obtaining ED drugs has opened the door for several counterfeiters to sell drugs online–wrongfully boasting of distribution of safe, efficacious ED drugs in bulk and for a fraction of the cost. In addition to online sales, many people have reported receiving the drugs illegally from "friends," dealers, and sex shops. Thought to be used mostly in older men suffering from erectile dysfunction, ED drugs are now abused by men and women of all ages, many of whom are using the drugs strictly for recreational purposes. Special populations at risk of abusing ED drugs include homosexual men, the sexually compulsive, college-aged individuals, those with cardiovascular disease, depression, or nicotine addiction. As more ED drugs are being developed to provide better results in less time, it will be extremely important to develop ways to control their distribution and ensure these drugs are being used solely for their medical purposes.

REFRENCES

1. Impotence. NIH Consens Statement Online. 1992; 10(4):1-31. Available at http://consensus.nih.gov/1992/1992Impotence091html.htm.

2. MIDASBLUE. Available at http://www.midusblue.co.uk/index.html.

3. Pfizer Labs. 69-5485-00-9: Viagra (sildenafil citrate) tablets. Pfizer Inc. NY, NY. 2003. Available at http://pfizer.com/pfizer/download/uspi_viagra.pdf.

4. Steers W, Guay AT, Leriche A, Gingell C, Hargreave TB, Wright PJ, Price DE, Feldman RA. Assessment of the efficacy and safety of Viagra (sildenafil citrate) in men with erectile dysfunction during long-term treatment. Int J Impot Res. 2001; 13(5): 261-267.

5. Paige NM, Hays RD, Litwin MS, Rajfer J, Shapiro MF. Improvement in emotional well-being and relationships of users of sildenafil. The Journal of Urology. 2001; 166(5):1774-1778.

6. Muller MJ, Benkert O. Lower self-reported depression in patients with erectile dysfunction after treatment with sildenafil. Journal of Affective Disorders. 2001; 66(2-3):255-261.

7. Nurnberg HG, Lauriello J, Hensley PL, Parker LM, Keith SJ. Sildenafil for sexual dysfunction in women taking antidepressants. Am J Psychiatry. 1999; 156(10): 1664.

8. Ng J, Finney SJ, Shulam R, Bellingan GJ, Singer M, Glynne PA. Treatment of pulmonary hypertension in the general adult intensive care unit: a role for oral sildenafil? British Journal of Anaesthesia. 2005; 94(6): 994-777.

9. Crohn's & Colitis Foundation of America. About Crohn's Disease. 2006. Available at http://ccfa.org/info/about/crohns.

10. Marks DJ, Harbord MW, MacAllister R, Rahman FZ, Young J, Al-Lazikani B, Lees W, Novelli M, Bloom S, Segal AW. Defective acute inflammation in Crohn's disease: a clinical investigation. Lancet. 2006; 367(9511): 668-678.

11. Akash R, Hrishikesh D, Amith P, Stafanous S. Association of combined nonarteritic anterior ischemic optic neuropathy (NAION) and obstruction of cilioretinal artery with overdose of Viagra. Journal of Ocular Pharmacology and Therapeutics. 2005; 21(4):315-317.

12. Pharmacia & Upjohn Company. Caverject impulse: dual chamber system alprostadil for injection. 2003. Available at http://www.caverject.com/CaverjectImpulseUSPISept2003.pdf.

13. Schwarz Pharma. Complete prescribing information: Edex (alprostadil for injection). 2004. Available at http://www.edex.com/EdexFullPI.pdf.

14. Erectile Dysfunction Information Center. Impotence treatments: urethral suppositories (MUSE(r)). 2005. Available at http://www.cure-ed.org/Erectile-Dysfunction-Treatment/Urethral-Suppositories.html.

15. Amorphine (Uprima) for the treatment of erectile dysfunction. MTRAC, Department of Medicines Management, Keele University. 2001. Available at http://www.keele.ac.uk/depts/mm/MTRAC/ProductInfo/summaries/A/APOMORPHINE2%202.pdf.

16. Bayer Pharmaceuticals Corporation. Prescribing information: Levitra (vardenafil HCl) tablets. 2005. Available at http://www.levitra.com/index.htm.

17. Lilly ICOS LLC. Prescribing information: Cialis (tadalafil) tablets. Eli Lilly and Company, Indianapolis, IN. 2005. Available at http://pi.lilly.com/us/cialis-pi.pdf.

18. Yohimbe. Whole HealthMD Supplements. 2000. Available at http://www.wholehealthmd.com/refshelf/substances_view/1,1525,830,00.html.

19. Yohimbe (Systemic). MedlinePlus. Microdex, Inc. 2000. Available at http://www.nlm.nih.gov/medlineplus/druginfo/uspdi/202639.html.

20. Johnston LD, O'Malley PM, Bachman JG, Schulenberg JE. Monitoring the Future national results on adolescent drug use: Overview of key findings, 2005. Bethesda MD: National Institute on Drug Abuse. 2006. Available at http://www.monitoringthefuture.org/pressreleases/05drugpr_complete.pdf.

21. Brenson A. Sales of impotence drugs fall, defying expectations. New York Times. New York Times Company. December 4, 2005.

22. Crosby R, DiClimente RJ. Sexual Behavior: Use of recreational Viagra among men having sex with men. Sex Transm Infect. 2004; 80(6):466-468.

23. Chu PL, McFarland W, Gibson S, Weide D, Henne J, Miller P, Partridge T, Schwarcz S. Viagra use in a community-recruited sample of men who have sex with men, San Francisco. Journal of Acquired Immune Deficiency Syndromes. 2003; 33(2):191-193.

24. Kim AA, Kent CK, Klausner JD. Increased risk of HIV and sexually transmitted disease transmission among gay or bisexual men who use Viagra, San Francisco 2000-2001. AIDS. 2002; 16(10):1425-1428.

25. Cachay E, Mar-Tang M, Matthews WC. Screening for potentially transmitting sexual risk behaviors, urethral sexually transmitted infection, and sildenafil use among males entering care for HIV infection. AIDS Patient Care STDS. 2004; 18(6):349-354.

26. Colfax GN, Mansergh G, Guzman R, Vittinghoff E, Marks G, Rader M, Buchbinder S. Drug use and sexual risk behavior among gay and bisexual men who attend circuit parties: a venue-based comparison. Journal of Acquired Immune Deficiency Syndromes. 2001; 28(4):373-379.

27. Delate T, Simmons VA, Motheral BR. Patterns of use of sildenafil among commercially insured adults in the United States: 1998-2002. International Journal of Impotence Research. 2004; 16(4):313-318.

28. Aldridge J, Measham F. Sildenafil (Viagra) is used as a recreational drug in England. BMJ. 1999; 318(7184):669.

29. Mondaini N, Ponchietti R, Muir GH, Montorsi F, Di Loro F, Lombardi G, Rizzo M. Sildenafil does not improve sexual function in men without erectile dysfunction but does reduce the postorgasmic refractory time. International Journal of Impotence Research. 2003; 15(3):225-228.

30. Polles A, Graham NA, Carnes SL, Carnes PJ, Gold MS. Use and abuse of erectile dysfunction drugs. American College of Clinical Pharmacology 35th Annual Meeting. Cambridge, Massachusetts. September 18, 2006.

doi:10.1300/J069v25S01_06

Index

Addiction treatment programs
    referrals to, 1
Alzheimer's disease, nicotine and, 24
Amphetamines, 7
Anabolic androgenic steroids (AAS), 6,9-10
    abuse of
        adverse effects of, 34-35
        and prevalence of psychiatric symptoms, 35-36
    addictive properties of, 36-37
    aggression and, 39-40
    classes of, 34
    effects of, 34
    increasing abuse of, 33-34
    mechanisms for effects of, 37-40
    original intent of, 34
    side effects of, 10
Anabolic Steroid Enforcement Act (1990), 9-10
Anesthesiologists, mortality rate of, 3
Anti-doping programs, 7
Attention, nicotine and, 23
Attention Deficit/Hyperactivity Disorder (ADHD),
    nicotine and, 23-24

Bell, Andrew, 62
Brown, David, 62
Byck, Robert, 2

CAVERJECT Impulse, 64
Cialis, 64
Cosmetic psychiatry, 1
Counterfeit drugs, problem of, 10
Crohn's disease, Sildenfil for treating, 63
Cushing, Harvey, 10

Designer steroids, 8
Diagnostic and Statistical Manual of Mental Disorders,
    fourth edition (DSM-IV), 2
Doping
    origin of word, 6
    in sports, 7
Drug abuse. See Performance-enhancement drug
    abuse; Steroid abuse
Drug seeking, performance-enhancing, 1
Dunn, Peter, 62

Ecstasy (MDMA). See MDMA (3-4 methylenedioxy-
    methamphetamine)
EDEX, 64
Ellis, Peter, 62
"Emperor's Clothes" phenomenon, 1-2
Erectile dysfunction (ED). See also Viagra
    alternative therapies for, 64-65
    causes of, 62
    defined, 61-62
    extent of in U.S., 62
    treatment options for, 62
Erythropoietin (EPO), 6,11-12

Feighner criteria, 2
Furchgott, Robert, 62

$GABA_A$ receptors, 37-38
Gay men, Viagra use and, 65-66

Hicks, Thomas, 7
Hoffman, Bob, 9
Homosexual men, Viagra use and, 65-66
Human growth hormone (hGH and rhGH), 6,10-11
Hypothalamic-Pituitary-Adrenal (HPA) axis, 38-39

Ignarro, Louis, 62
Impotence. See Erectile dysfunction (ED)

Johnson, Ben, 7,11

Levitra, 64
Linton, Arthur, 6

MDMA (3-4 methylenedioxy-methamphetamine)
    addictiveness of, 52-53
    consumption of, 48
    drug classification of, 49
    emergency room cases and, 49
    history of, 48
    increase in use of, 48-49
    long-term chronic effects of, 52
    medical consequences of, 51-53

metabolism of, 50
neurotoxicity associated with, 53-54
overview of, 48
pharmacology of, 49-50
research studies on, 54
as social facilitator, 50-51
tablet form, 48
Memory, nicotine and, 21-23
Monoaminergic systems, 39-40
Murad, Ferid, 62
MUSE (medicated urethral system for erection), 64

Neuronal nicotinic acetylcholine receptors (nAChRs), 18-20
types of states of, 20-21
Neuropeptides, 38
Nicotine
Alzheimer's disease and, 24
attention and, 23
Attention Deficit/Hyperactivity Disorder and, 23-24
behavioral effects of, 21
effects of, 17-18
memory and, 21-23
Parkinson's disease and, 24
Nicotine receptor. *See* Neuronal nicotinic acetylcholine receptors (nAChRs)
Nicotine replacement therapy (NRT), 25-28

Olympic Games
doping in, 7
performance enhancement in, 6

Parkinson's disease, nicotine and, 24
Performance enhancement
methods of, 6
in Olympic Games, 6
Performance-enhancement drug abuse. *See also* Steroid abuse
at-risk populations for, 12-13
growing use of, 13-14
Performance-enhancer users, 3
Performance-enhancing drugs
categories of, 8-9
historical overview of, 6-8
popular, 8-9
Performance-enhancing drug seeking, 1
Prescription medication use, 3
Psychiatric diagnoses, 1-3

Research Diagnostic Criteria, 2

Sex addicts, Viagra use and, 66
Sildenafil, development of, 62. *See also* Viagra
Smoking. *See* Nicotine
Sports
anti-doping programs in, 7-8
doping in, 7
Steroid abuse, 1. *See also* Performance-enhancement drug abuse
at-risk populations for, 13
substance abuse and, 13
Steroid abuse scandals, 7
Steroids, designer, 8. *See also* Anabolic androgenic steroids (AAS)
Substance P (SP), 38

Terrett, Nicholas, 62
3-4 methylenedioxy-methamphetamine (MDMA). *See* MDMA (3-4 methylenedioxy-methamphetamine)
Tobacco, 17-18. *See also* Nicotine
addictive nature of, 24-25
treatment for addiction to, 25

UPRIMA, 64

Viagra. *See also* Erectile dysfunction (ED)
adverse effects of, 63-64
development of, 62-64
imitations of, 65
non-prescription use of, 65-66
novel uses for, 63
prescribing information for, 62
success of, 62-63
use of
by college-aged individuals, 66
by compulsive sex addicts, 66
by homosexual men, 65-66

Wood, Albert, 62
World Anti-Doping Agency (WADA), 7
categories of prohibited substances of, 8-9

Yohimbe, 64

Ziegler, John, 9